Wagar Media Center
11270 Grafton Road
Carleton, MI 48117

MODERN WORLD NATIONS

AFGHANISTAN	IRAQ
ARGENTINA	IRELAND
AUSTRALIA	ISRAEL
AUSTRIA	ITALY
BAHRAIN	JAMAICA
BERMUDA	JAPAN
BOLIVIA	KAZAKHSTAN
BOSNIA AND HERZEGOVINA	KENYA
BRAZIL	KUWAIT
CANADA	MEXICO
CHILE	THE NETHERLANDS
CHINA	NEW ZEALAND
COSTA RICA	NIGERIA
CROATIA	NORTH KOREA
CUBA	NORWAY
EGYPT	PAKISTAN
ENGLAND	PERU
ETHIOPIA	RUSSIA
FRANCE	SAUDI ARABIA
REPUBLIC OF GEORGIA	SCOTLAND
GERMANY	SOUTH AFRICA
GHANA	SOUTH KOREA
GUATEMALA	TAIWAN
ICELAND	TURKEY
INDIA	UKRAINE
IRAN	UZBEKISTAN

Bosnia and Herzegovina

Douglas A. Phillips

Series Consulting Editor
Charles F. Gritzner
South Dakota State University

An imprint of Infobase Publishing

This book is dedicated to the heroic educators in Bosnia and Herzegovina and in the United States who have worked together with the Center for Civic Education to advance democratic education in Bosnia-Herzegovina and the Balkans. The brave civic educators from Bosnia and Herzegovina and their talented students are working to shape an incredible democratic future for that country.

Frontispiece: Flag of Bosnia and Herzegovina

Cover: City of Mostar, shown in 1976, when its famous bridge, the Stari Most bridge, had not yet been destroyed in the 1992–1995 war.

Bosnia and Herzegovina

Copyright © 2004 by Infobase Publishing

All rights reserved. No part of this book may be reproduced or utilized in any form or by any means, electronic or mechanical, including photocopying, recording, or by any information storage or retrieval systems, without permission in writing from the publisher. For information contact:

Chelsea House
An imprint of Infobase Publishing
132 West 31st Street
New York NY 10001

Library of Congress Cataloging-in-Publication Data applied for.

ISBN 0-7910-7911-2

Chelsea House books are available at special discounts when purchased in bulk quantities for businesses, associations, institutions, or sales promotions. Please call our Special Sales Department in New York at (212) 967-8800 or (800) 322-8755.

You can find Chelsea House on the World Wide Web at http://www.chelseahouse.com

Text design by Takeshi Takahashi
Cover design by Takeshi Takahashi, Keith Trego

Printed in the United States of America

Bang 21C 10 9 8 7 6 5 4 3 2

This book is printed on acid-free paper.

Table of Contents

1	Introducing Bosnia and Herzegovina	8
2	Physical Geography	16
3	Bosnia and Herzegovina Through Time	32
4	Disintegration and War	42
5	People and Culture	52
6	Government and Politics	64
7	Economy	74
8	Cities	84
9	Bosnia and Herzegovina Looks Ahead	94
	Facts at a Glance	102
	History at a Glance	104
	Further Reading	106
	Index	107

Bosnia and Herzegovina

CHAPTER 1

Introducing Bosnia and Herzegovina

Time seems to move slowly in Bosnia and Herzegovina. This is evident in the regular stroll, called a *korzo*, which takes place in Sarajevo and numerous other towns and villages each evening. Families and friends stroll at a casual pace, pausing to greet people they know. Laughter and storytelling mix with background music. Coffee and food shops tempt the senses and line the streets where the korzo takes place. Teens taking the korzo look for friends or perhaps, for that special young man or woman they would like to know better. The korzo is a nightly social event, without parallel in the West. It is something special, something to be remembered and celebrated.

Even the rolling hills of Bosnia and Herzegovina's beautiful countryside seem to reflect the slower pace of time in this region. This pace of life can be deceiving though, as change has often come too rapidly to this new, small, and often troubled country. On a nondescript bridge in Sarajevo, for example, the flames igniting World War I were kindled. It was on this bridge that Archduke Franz Ferdinand and his wife, Sophie, were assassinated on an ill-fated day in 1914. This event, in the seemingly calm Sarajevo, triggered the chain of events that resulted in World War I.

Today, Bosnia and Herzegovina is a war-tattered remnant of the former Yugoslavia, which began to dissolve in the early 1990s. Five countries ultimately emerged from the breakup: Bosnia and Herzegovina, Croatia, Macedonia, Slovenia and, in 2003, Serbia and Montenegro. Why was Yugoslavia divided? A commonly accepted reason is religious differences. This offers a partial answer. The real answer is much more complex. It lies in deeply rooted ethnic differences, a strong sense of national identity among different peoples, and numerous other forces. These elements combined to bring about the demise of Yugoslavia, and the complicated birth of Bosnia and Herzegovina and the other newly emerged countries.

Bosnia and Herzegovina has three primary religious groups: Orthodox, Catholic, and Islam. Orthodox Christians have ties to the Serb ethnic group, Catholics to the Croats, and the Muslims are often referred to as *Bosniaks* (not to be confused with Bosnians). Surprisingly, each of these ethnic groups has the same ancestral heritage, just as the now diverse population shares the same Slav ethnic roots. Why, then, the tragic divisions? Religious differences and extreme nationalism offer the most apparent answers to this question. This complex question will be examined in greater detail in the following chapters, and it is an essential one to explore if one is to understand Bosnia and Herzegovina and its neighbors in the region.

A visitor arriving in the capital city of Sarajevo is immediately awestruck by the beauty of this small country. Rolling hills dominate the landscape and form a picturesque frame around the city. Yet these beautiful hills can be deceiving. During the war in the early 1990s, they held horror and constant threat for the local population. In these hills, Serb gunmen hid, and often fired on people dashing in and out of their homes and businesses. The gunmen fired randomly at helpless citizens, in a sadistic activity that massacred thousands.

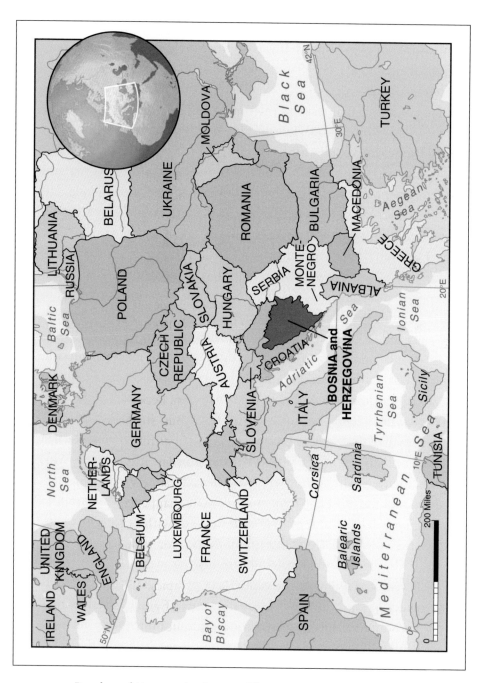

Bosnia and Herzegovina is one of five countries that resulted from the breakup of the former Yugoslavia. Serbia and Montenegro are located to the east and southeast while Croatia lies to the north and west.

Introducing Bosnia and Herzegovina 11

The area near this Holiday Inn became known as the deadly shooting gallery, "Sniper's Alley." During the war in the early 1990s, Serb gunmen hid in the hills around Sarajevo and killed people by the thousands.

One area, near the vividly yellow Holiday Inn, became known as "Sniper's Alley," because of the dangerous strip of land between the hotel and local businesses and apartments, where many Sarajevo residents met their death. Today, it is hard to believe that a person could be killed while simply running across the grassy strip to buy a pizza, yet it happened. In the hills surrounding Sarajevo, danger lurked, with both large and small weapons aimed at the city as it lay for years under a devastating siege.

Sarajevo, host to the 1984 Winter Olympics, still retains sadness associated with the war. Thousands of graves now lie beside the Olympic skating venue; gravestones dot the landscape, marking the war's terrible toll. Local residents call this area the Children's Cemetery. The scale of the cemeteries and magnitude of human loss always strikes residents and visitors alike with a deep sense of sadness. Reading the tombstones, the visitor will see the old and young, men and women, Croat,

The 1984 Winter Olympics were a high point in Sarajevo history, but now thousands of graves lie beside the skating venue that was used during the Olympic games.

Bosniak, and Serb all buried near each other in a stark reminder of the atrocities and horror of the war. Bullet holes and bomb scars still mar the landscape of Sarajevo and other cities, but rebuilding and healing are also evident. There is much more to the story of this country, however, than bullet and mortar-scarred buildings and tales of horror and atrocity.

Bosnia and Herzegovina is a triangular country with an area of 19,741 square miles (51,129 square kilometers), slightly smaller than West Virginia. The country's shape is symbolized on its flag, which bears a yellow triangle. The population of Bosnia and Herzegovina is just under four million, more than twice that of West Virginia. This number is far lower than before the war, as hundreds of thousands were killed, or fled to safer countries. Former enemies surround the country. Serbia and Montenegro are located to the east and southeast while Croatia is wrapped around the country to the north and west. Relations with these neighbors have improved, but difficulties still persist. Sarajevo is not only the country's capital, it also is the largest city, with a population of about 400,000. Banja Luka and Mostar are the country's second and third-largest cities.

Politically, Bosnia and Herzegovina is composed of two entities. An *entity* is similar to a state in the United States, or a province in Canada. The northern, primarily Serb, entity within the country is called the Republic of Srpska, with its capital in Banja Luka. The Federation is the second entity of the country. This entity has all three ethnic groups present, and it is formally referred to as the Federation of Bosnia and Herzegovina. Both the Republic of Srpska and the Federation of Bosnia and Herzegovina are entities within the country, which is itself called Bosnia and Herzegovina. This can be confusing, but it is a bit like the city of Washington, D.C. and the state of Washington. The name used for both is similar, but refers to either the city or the state. In the country of Bosnia and Herzegovina, "Federation" is the term normally used by residents when referring to the entity (province) of Bosnia and Herzegovina.

Driving through the forested countryside of northern Bosnia and Herzegovina, rolling hills greet the visitor at every bend in the road. The country extends from the humid mountainous north, where snow falls in the winter, to the southern locations like the city of Mostar, where a dry, almost desert-like climate is found. Forests dominate the land in the north, while the land becomes flatter and more agricultural in the south. This southeastern region is called Herzegovina. Bosnia and Herzegovina even has a narrow stretch of coastline along the Adriatic Sea. Squeezed between Croatia, and Serbia and Montenegro, the coastline is of limited use as it is only thirteen miles long. Some opportunity may ultimately develop, however, to create connections for trade and sea travel, even though no natural harbors now exist on that coastline.

The natural landscape of Bosnia and Herzegovina is very attractive. However, it is the country's people that are the real treasure. With a hospitality that is unmatched, the Bosnians greet visitors to their homes with a smile and small, yet very strong cups of coffee that are sure to keep one awake for hours. The coffee is accompanied by talk of the events of the day, music, family, basketball or soccer, politics, and other issues of importance. Stories of the war's horrors are less frequent now than during the late 1990s, but the scars are deep in the hearts of Bosnians. Nearly everyone lost either family or friends or both, or had people close to them injured during the hostilities.

The unsavory term "ethnic cleansing" is synonymous with the war in Bosnia and Herzegovina. This seemingly sanitary term was used for such incredibly inhumane activities as torture, slavery, removal, concentration camps, and death for thousands of Serbs, Croats, Bosniaks and others lost in the war. Nearly one million people fled or were forced from their homes by other ethnic groups. These people lost their homes and all of their possessions, and became known as displaced people.

Unlike its neighbors Croatia and Serbia and Montenegro, Bosnia and Herzegovina has a diverse population, which is made up of 40 percent Serbs, 38 percent Muslims, and 22 percent Croatians. By contrast, Croatia is 78 percent Croatian and Serbia is 65 percent Serb. The rate of intermarriage between religious and ethnic groups in Bosnia and Herzegovina was very high, especially in the cosmopolitan city of Sarajevo, where nearly 30 percent of the married population were in religiously and ethnically mixed marriages. This diversity within Bosnia and Herzegovina made the war very complex.

The war in the 1990s is the recent legacy of Bosnia and Herzegovina. What about the present and the future? This country has a well-educated population that is ready to work. Because of lingering perceptions about the war, jobs are still difficult to find. Yet, the potential for tourism and manufacturing is very evident. Bosnians joke that before the war they only spoke one language, Serbo-Croatian, but now they speak three. Because the war divided the peoples, three languages were claimed by different nationalist entities. Thus, all Serbo-Croatian speakers became the speakers of Croatian, Serbian, and Bosnian languages. In reality, all three are still the same Serbo-Croatian language.

Politically, Bosnia and Herzegovina has as an extremely complex governing process. The peace talks ending the war in the Balkans were held in Dayton, Ohio. Thus, the resulting agreement is called the Dayton Accord. This document sets forth the constitution and the political structure that was then externally imposed upon Bosnia and Herzegovina. With the Dayton Accord, the country has a democratic constitution that recognizes the major ethnic divisions within the country's political structure. This aspect of the constitution has made political decision-making very awkward and sometimes impossible.

In the distant past Bosnia and Herzegovina served as a crossroad for powerful neighbors, ranging from the Romans in the first century A.D., to Hungarians in the twelfth century, and Ottoman Turks in the fifteenth century. Its location has made it a route for warriors moving between Asia-Minor (peninsular, or Asiatic, Turkey), Greece, and other parts of Europe. In the twentieth century, both the Nazis and Communists had control of this volatile region of the world. A location at the crossroads can provide huge disadvantages, but it also has positive aspects with respect to trade and can serve as a place where people exchange important ideas. The importance of this region's location has been demonstrated on numerous occasions.

This book takes you on a stimulating korzo through Bosnia and Herzegovina, both past and present. As on the evening stroll, you will gaze upon the people, places, and events that have placed Bosnia and Herzegovina so prominently on the world's stage. Through your journey, you will gain a better understanding of, and appreciation for, this small, but amazing country nestled away in the heart of the Balkans.

CHAPTER 2

Physical Geography

The physical geography of Bosnia and Herzegovina is perhaps most noteworthy for its lack of extreme conditions. In terms of its physical geography, little of the country stands out as being unique. Its location, squarely in the midlatitudes, ensures a climate lacking extreme temperatures. The country's land features include small areas of plains and river valleys, but rugged terrain dominates the landscape. Although there are several important rivers, surface water is limited by the country's very porous rock material. Officially, in fact, Bosnia and Herzegovina has no surface water. Bosnia and Herzegovina does border the Adriatic Sea, but only claims about 13 miles (21 kilometers) of coastline on this arm of the Mediterranean. Plant and animal life have been severely altered by thousands of years of human habitation. Nonetheless, the country does have areas in which subtropical crops can be grown throughout the year. In contrast, it also has hosted the Winter Olympics! In this chapter, you will learn about

Bosnia and Herzegovina is a rugged land filled with hills, plateaus, and mountains. Numerous high valleys are sandwiched between towering peaks.

the physical environments and landscapes, and their importance to the residents of Bosnia and Herzegovina.

LOCATION

Bosnia and Herzegovina is situated squarely in the middle latitudes—halfway between the Equator and the North Pole. The 45th parallel of latitude cuts across the northern, and widest, part of this triangular country. Geographically, Bosnia and Herzegovina is clustered with those states collectively referred to as being southeast European in their location. Additionally, many geographers use the term "Balkan Peninsula" to describe the lands south of the Sava and Danube rivers. From north to south, this division includes: Bulgaria (and occasionally Hungary and Romania), Slovenia, Croatia, Serbia and Montenegro, Macedonia, Albania, European Turkey, Greece, and, of course, Bosnia and Herzegovina. The regional term, itself, is derived from the Balkan Mountains, an east to west trending range in Bulgaria and Serbia and Montenegro. Most people living in the region prefer to be referred to as Southeast Europeans, rather than "Balkanians."

Interestingly enough, Bosnia and Herzegovina owes its present existence and location to historical circumstances of nineteenth-century geopolitics. Were it not for the independence movements in neighboring countries that left this Ottoman province in the hands of the Habsburg monarchy, and later the former Yugoslavia, the territory of Bosnia and Herzegovina would perhaps be an integral part of Croatia or Serbia and Montenegro. Today, a particular ethnic group identifies most countries in the region; for example, Turks in Turkey, and Greeks in Greece. There are, however, no such ethnic groups as Bosnians or Herzegovinans.

Major rivers and mountain ranges determine Bosnia and Herzegovina's boundaries. In the north, the boundary is shared with Croatia, much of which follows the Sava and Una rivers. The country's eastern boundary with Serbia and Montenegro follows the Drina River. The Dinaric Alps (so named from Dinara Mountain, a peak in the range) forms both a natural and administrative western boundary with Croatia. At the city of Neum, on the Adriatic coast, Bosnia and Herzegovina has its only narrow face on the Adriatic Sea, making the country for all practical purposes landlocked. The value of this seacoast is very limited. It is but a thin sliver of land, almost completely surrounded by Croatian territory. The southern border between Serbia and Montenegro follows the old boundary defined in the former Yugoslavia after World War II.

LAND FEATURES

Geologically, Bosnia and Herzegovina, for the most part, share similar past and present-day landform features with most areas of the Balkan Peninsula.

Physical Geography 19

Major rivers and mountain ranges determine Bosnia and Herzegovina's borders. It has only a very narrow window on the Adriatic Sea and is, for all purposes, landlocked.

Geologic Conditions

You might wonder why a country's geologic past is important to its geographic present. There are several reasons why geology often plays an important role in developing a country's geographical conditions. In Bosnia and Herzegovina, for example, *karst* features are commonplace. The limestone-formed hills and plateaus have been deeply eroded by rain, leaving a rugged terrain etched into grooves, terraces, valleys, and caves. Such karst features exist only in certain types of rock, the nature of which is the result of geologic processes acting, in some cases, millions of years ago. The country's mountains are the result of geologic processes working through time, often over millions of years. Mineral deposits, too, are the result of various geologic processes. Additionally, severe earthquakes occasionally rock the region, again, as a result of geologic processes operating from deep within Earth's crust.

Most highlands of Bosnia and Herzegovina are geologically considered to be a continuation of a calcareous belt of mountains called the Southern Alps. The Dinaric Alps, the country's highest range, which forms the western border with Croatia, is composed mainly of ancient limestone and dolomite rock. When exposed to chemical action (for example, exposure to water), limestone and dolomite are readily eaten away. The results of this chemical weathering are a variety of distinct landform features that collectively are called *karst*.

Most of Bosnia and Herzegovina can best be described as "rugged." Hills, plateaus, and low mountains dominate the landscape. The forces of weathering (rock decomposition and disintegration) also have contributed to landscape diversity, having sculpted and deposited a variety of unique features.

A Rugged Land

In Bosnia and Herzegovina, travelers seeking information about travel between two points will almost always ask about travel time, rather than distance. This is because most of the

Physical Geography 21

This picture of Sarajevo in winter shows the peaks around the city that define the rugged western half of the country.

country's terrain is rugged hills, plateaus, and mountains. Here, distance means little. When traveling in many areas of the country, it seems as though more time is spent going up and down than toward one's destination!

The rugged backbone of the Dinaric Alps covers, roughly, the western half of Bosnia and Herzegovina. Near the cities of Travnik, Sarajevo, and Mostar, peaks rise to elevations exceeding 6,000 feet (1,829 meters), reaching 7,831 feet (2,387 meters) atop Mount Maglic, along the southern border with Serbia and Montenegro. Numerous high valleys are sandwiched between towering peaks. The most noteworthy are the plains of Livno and Glamoc, which lie parallel to Dalmatia (the region of coastal Croatia bordering Bosnia and Herzegovina), at an elevation of 500 feet (152 meters).

A region of moor and forest-covered mountains, plateaus, and hills with deeply entrenched alluvial basins extends to the northeast of the Dinaric Alps. Because of its rugged terrain, this is the least densely populated area in the country. Southern Bosnia and Herzegovina has some of the country's most unique land features. One interesting aspect to this karst area is the formation of rivers that appear from a cliff face, disappear underground, and reappear some distance away. These areas are generally infertile and rather unsuited to extensive settlements due to their jagged landscape, caves, and dry, porous soils.

Lowlands

A land-use map of Bosnia and Herzegovina shows that widespread mixed farming is practiced only in the northern part of the country. Here, a narrow band of relatively flat, fertile plains—no more than 50 to 100 miles (80-160 kilometers) in width—supports an extensive industry of crops and livestock. Not surprisingly, it also is the most densely populated section of the country. Soils, such as those in Egypt's fertile Nile Valley, are rich because they are alluvial in origin. That is, they have been deposited and repeatedly renewed through flooding of the Sava River and its tributaries, over thousands of years. Other than for a few narrow river-valley plains, the only other relatively flat land is in the southwest, where alluvial deposits from the Neretva River have formed a small floodplain.

WEATHER AND CLIMATE

Weather is defined as the current condition of the atmosphere; *climate* is a long-term average of weather conditions. Both are influenced by a combination of controls, including latitude; proximity (closeness) to a large body of water; and elevation. As previously mentioned, Bosnia and Herzegovina is located exactly halfway between the Equator and the North Pole, placing the country within the midlatitudes. Its location, approximately 43 to 45 degrees north latitude, places it at the same distance from the equator as Oregon and southern Minnesota. However, because of its closeness to the Mediterranean Sea, the country enjoys relatively warmer winters, and somewhat cooler summers than do other areas at similar latitude that are located farther inland. Generally, the temperatures of Bosnia and Herzegovina are much more similar to coastal Oregon than to those of Minnesota, which is located in the heartland of the United States. Throughout most of Bosnia and Herzegovina, elevation plays an important role in temperature.

Most of the country experiences slightly warmer summers and cooler winters than do many areas of Mediterranean climate. To understand why this occurs, one must think in terms of that primary geographical element—location—and also such environmental conditions as prevailing wind direction, proximity to a large body of water, and mountain barriers. In Oregon, huge differences exist in both temperature and moisture between the western, coastal part of the state and the interior, lying east of the Cascade Mountains. In the case of Bosnia and Herzegovina, even though it is very close to the Adriatic Sea, the Dinaric Alps somewhat block the country from the sea's moderating affects. Bosnia and Herzegovina has three climatic types: Mediterranean in the west and southwest; Humid Subtropical in the north and east; and Highland conditions in mountainous areas.

Mediterranean Climate

Much of western Bosnia and Herzegovina has a Mediterranean climate and ecosystem—very similar to conditions found along much of the West Coast of the United States, west of the Sierra Nevada and the Cascade Mountains. Many people believe the Mediterranean to be the world's most pleasant climate. In addition to a lack of extreme temperatures, there are moderate, yet quite recognizable, seasonal changes. The land also receives ample precipitation, although it is unevenly distributed throughout the year. In fact, perhaps the most unique characteristic of the Mediterranean climate is the long period of high-sun season (summer) drought.

The Mediterranean is a sunny climate, averaging 2,500 hours of sunlight yearly. Although the western region averages about 35 inches (89 centimeters) of precipitation each year, it has a marked seasonal distribution. The summer, which throughout most of the world is the wettest season, is quite dry. Winter is the season of moisture, some of which can fall as snow throughout much of the country, particularly at higher elevations. Although the region typically experiences quite moderate temperatures, the country's highest recorded temperatures occur in this climatic region, not far from the Croatian border. As a matter of fact, it is one of the "hottest" places in the former Yugoslavia. Temperatures can be sweltering, climbing to more than 100° F (38° C) during the summer. Throughout the winter months, temperatures can plunge to well below freezing. In July and August, afternoon temperatures climb to a reasonably comfortable average of 75 to 85° F (24 to 30° C). During December and January, nighttime temperatures drop to an average 25 to 35° F (-4 to 2° C). Summer highs and winter lows vary somewhat, depending upon elevation. In the Neretva Valley, mandarin oranges grow, while skiers enjoy their sport at high mountain elevations.

Humid Subtropical Climate

Limited areas of eastern Bosnia and Herzegovina, particularly those lying at lower elevations, such as the northeastern lowland plains, experience a Humid Subtropical climate, similar to that of the southeastern United States. Summers range from warm to hot and quite humid. A long growing season and adequate precipitation spread throughout the year help make this the country's most productive agricultural area. Winters are cool and also quite moist. Occasional blasts of frigid Siberian air settle over the region, bringing a sharp drop in temperature, and often last for several weeks. The high-pressure system brings crystal clear skies and no precipitation.

Mountain Climates

On average, temperatures drop about 3.5° F (2° C) with each 1,000-foot (305-meter) increase in elevation. With much of Bosnia and Herzegovina situated upland, elevation is perhaps, the single most important control of temperature. It also plays an important role in precipitation.

In higher mountain regions, summers are short and cool. Winters are long and cold, and can bring many feet of snow that can linger well into the late spring. Temperatures can plunge well below 0° F (-18° C). The lowest temperatures throughout the Balkan Peninsula coincide with the arrival of occasional continental polar air masses that bring bone-chilling cold.

Highlands also influence the distribution of precipitation. The windward side of a mountain receives much more moisture than does the leeward (downwind) side. Westward-facing mountain slopes in northern California and Oregon, for example, may receive upwards of 100 inches (254 centimeters) of precipitation. On the leeward side of the Cascades, just a short distance away, the landscape is one of parched desert. In the Balkan Peninsula, the prevailing wind direction is from the west. As the moisture-bearing winds from the Adriatic Sea blow upslope, precipitation falls on the windward, or western, side of the Dinaric Alps. Here, more than 60 inches (152 centimeters) of moisture is received each year. On the leeward (eastern) side of the mountains, some areas receive as little as 20 inches (51 centimeters) of precipitation. The combination of lower temperatures and greater moisture contributes to heavy snowfall throughout the winter that, on higher peaks, lingers well into the spring.

PLANT AND ANIMAL LIFE

A Sarajevo museum has a collection of local flora, or plant life, that includes more than 3,000 species, some of which are rare. The country is also home to a wide variety of fauna, or animal life.

"Bosnia begins with the forest" is a saying borne out by such towns as Vranduk, where the countryside is filled with dozens of species of trees. But the absence of responsible environmental planning and practices has led to the overcutting of trees and other abuses.

Plant Life

About half of Bosnia and Herzegovina is covered with forest. "Bosnia begins with the forest," says a native proverb, "Herzegovina with the rock." Broadly speaking, this is accurate. An exception is the karst landscape, which is quite bare of vegetation wherever it occurs in the country.

On mountain crests, only the hardiest lichens and mosses can survive. Below, lies a belt of woodland. Here, large timber includes many giant trees, some of which reach some 200 feet (61 meters) toward the heavens, and are 20 feet (6 meters) in girth at the level of a man's shoulder. Dense brushwood litters the foothills. In most lowland areas, crops and other domesticated plants replaced the natural vegetation centuries ago.

The largest, and most important, of the three zones is the woodland, which can be further divided into three subzones. At lower elevations, up to about 2,500 feet (762 meters) in the north, sunnier slopes are overgrown by oaks, and the shadier slopes by beeches. Farther south, in the central part of the country, oak begins to be replaced by beech, elm, ash, fir and pine, at elevations up to 5,000 feet (1,524 meters). The third zone, reaching to 6,000 feet (1,829 meters), is characterized by the predominance of fir, pine, and other conifers. Chestnut, aspen, willow, birch, alder, juniper, and yew occur in all three zones. Mountain ash, hazel, wild plum, wild pear, and other wild fruit trees are found scattered about the woodlands, particularly at lower elevations.

Until 1878, the forestlands were almost neglected. In that year, the government was forced to take action. It levied a graduated tax on goats, owing to the tremendous damage the animals inflicted upon young trees. It also curtailed the popular rights of cutting trees for timber and firewood, and of free pasturage for livestock. These measures were largely successful, although in 1902, the export of oak staves (wooden strips) was also discontinued, owing to short supply. Under Communist rule (1945–1990), illegal cutting by ordinary people was controlled relatively well, although corrupt local officials often allowed "legal" cutting by companies that did not follow environmental rules. During the war years there were no controls and clear-cutting continued, both by ordinary people and elected officials. Today, many forests are full of land mines, which are not only dangerous for people and animals, but also make the woodlands economically unproductive.

Animal Life

In 1893, the bones of a cave bear (*Ursus spelaeus*) were taken from a cave located in the Bjelasnica Range. This was a discovery without parallel in the Balkan Peninsula. Of existing species, the bear, wild boar, badger, roe deer, and chamois (mountain "goat" or antelope) are occasionally seen in the most remote mountainous and forested areas. Hares are uncommon, and the last red deer was shot in 1814; but wolves, otters, and squirrels abound. Large numbers of snipe, woodcock, ducks, and rails (wading birds) flock to the banks of the Drina and Sava Rivers. Cranes, pelicans, wild swans, and wild geese also are fairly plentiful. Several varieties of eagles and falcons still survive. For centuries, falconry was a popular pastime of the Muslim landlords.

Other forms of wildlife include two species of venomous snakes, the long-nosed viper and European adder. Karstlands are home to lizards and scorpions, some of which are venomous. The karst region also is home to some rather strange life forms. Caves and underground streams are inhabited by several curious kinds of fish, some of which are unknown elsewhere. Surface rivers teem with edible fish, including trout and salmon. Fishing has long been a popular sport, and it also is an important source of protein for many rural people. Eel fisheries on the Neretva River are of commercial value, as is leech gathering, one of the country's more unique industries.

During the period of Turkish rule, little attention was given to the preservation of wildlife. Widespread killing of wild game was commonplace. As a result, during the late nineteenth century, a series of laws were written to protect animals from needless slaughter. Among the more important laws were those limiting hunting to fixed seasons and requiring hunting licenses.

WATER FEATURES

Water features in Bosnia and Herzegovina can, for the most part, be explained in two terms: relief and karst. An estimated 90 percent of all the world's lakes are of glacial origin. Here, only a few small tarns—mountain lakes scoured out by Ice Age glaciers—exist at higher elevations. In the karst country, characterized by very porous limestone and dolomite rock, surface streams are few and underground streams prevail. In fact, it

The Neretva River is the only stream of any size flowing across the karst landscape. It flows past Mostar in southern Bosnia and Herzegovina, where minarets rise from a beautiful mosque near its banks.

is not at all uncommon for a stream to appear as a spring, disappear beneath the ground, and then emerge again.

The country's most important surface water is limited to six major rivers. Five of the streams flow northward, from the interior highlands. They follow the natural slope that gradually descends in elevation, toward the Sava River, a tributary to the Danube. From west to east, the main rivers are: the Una and its tributary, the Sana; the Vrbas, which flows through Banja Luka; the Bosna; and finally the Drina, which forms much of the country's border with Serbia and Montenegro. The Sava is navigable upstream to the Croatian city of Sisak, thereby being navigable throughout its course in Bosnia and Herzegovina.

Although the Sava does not have potential for hydroelectric development, several of the country's rivers do, including the Drina, Vrbas, and Neretva. The only stream of any size flowing across the karst landscape is the Neretva. This river, located in southern Bosnia and Herzegovina, flows northwestward, from the highlands near the border with Serbia and Montenegro; then reverses its direction to flow southwestward, through a narrow sliver of Croatia, and on into the Adriatic Sea.

Bosnia and Herzegovina has only a narrow window on the Adriatic Sea—about 13 miles (21 kilometers). In traveling from Mostar to Dubrovnik (Croatia) using a coastal route, one passes through Neum, the country's only town located on the sea. The city is a major tourist center on the eastern shore of the Adriatic.

MINERAL RESOURCES

Bosnia is rich in minerals, including coal, iron, and copper. It also has valuable deposits of chrome, manganese, cinnabar, zinc, and mercury. The country also produces excellent marble and building stone. In the mountains, gold and silver were mined during Roman times, two thousand years ago. Salt is obtained from pits at Tuzla. Mineral springs also abound, particularly in the central and northern areas of the country. Springs located near Sarajevo have been used as medicinal baths since the days of the Romans. During the early twentieth century, their popularity increased greatly, as people were attracted by the supposed medicinal properties of the spas.

ENVIRONMENTAL HAZARDS AND ISSUES

Bosnia and Herzegovina is relatively free of highly destructive environmental hazards. The country does suffer from occasional earthquakes, and both floods and wildfires can cause local damage.

Although the country does not suffer extensively from the natural elements, it does experience a number of environmental problems. These include air pollution (particularly from metallurgical plants), occasional water shortages, and a limited number of sites available for urban waste disposal. A number of the country's most severe problems stem from the civil conflicts of the 1990s. Land mines present an omnipresent threat to life and limb in many rural areas. During the war, huge areas of old-growth forest—centuries-old stands of oak, pine, beech, and other commercially valuable species—were clear-cut by corrupt profiteers, who sold the wood to Western countries at huge profits.

Citizens of Bosnia and Herzegovina are now developing a stronger sense of concern, ethics, and responsibility for the environment. This has not always been true, especially of rural people who in the past may have considered a river to be little more than a flowing garbage dump. Today regulations that protect the environment are increasing but problems of dumping, poaching, and pollution continue in some regions.

In the absence of responsible environmental planning and effective laws, both rural and urban areas suffer in many ways. The result is the gradual degradation of environmental quality—from overcutting of trees, to polluted air and streams, and the fouling caused by inappropriate waste disposal. Deteriorating water quality poses yet another problem, particularly water used for drinking and other domestic purposes. In karst areas, sewage and other substances seep down from the surface and easily contaminate underground water. Even the country's wildlife is endangered because of illegal poaching and loss of habitat.

CHAPTER

3

Bosnia and Herzegovina Through Time

The Sarajevo rose presents a remarkable visual twist. This "plant" blooms year-round, and its pinkish-colored petals can measure three to four feet, or one meter, across. Like Bosnia and Herzegovina, the flower represents something other than its actual appearance. The Sarajevo rose can be found throughout the city, even along sidewalks and in the old section called Starigrad. The rose also serves as a reminder of Sarajevo's recent past.

THE IMPORTANCE OF LOCATION

The history of Bosnia and Herzegovina is filled with as many twists and turns as the world's scariest roller coaster. The country's geographic location on the Balkan Peninsula has placed it in within a region that is viewed as being little more than a doormat to more powerful European and Asian countries and cultures. Others have trampled upon Bosnia and Herzegovina many times. This history of

After the war people put a rosy pink plastic substance into the holes blasted into the concrete by mortar rounds. These have come to be known as "Sarajevo roses," because of the flowerlike patterns of the holes. "Sarajevo roses" remain today as a symbol of the legacy of the war and its many victims.

oppression dates back to the first-century Romans, whose presence is evident, even today, on the landscape. Few people realize that the former Yugoslavia, of which Bosnia and Herzegovina was a part, had a wonderful Roman coliseum in the city of Pula (now located in Croatia). Many roads and bridges are just remnants of the Roman era that remain on the country's contemporary landscape.

Being a doormat at history's doorstep appears to be an unenviable position. Everyone steps on the doormat, leaving footprints, dust, and dirt. However, this experience can also serve as a wellspring of positive information for a culture. When a country is located at such a crossroads, many interesting ideas, including new inventions and innovations, move across its threshold. Positioned between many historically powerful cultures, Bosnia and Herzegovina has been a corridor for powerful armies moving between Asia Minor, Greece, Italy, and other parts of Europe. This chapter examines the cultural crossroads that have played a major role in the development and shaping of Bosnia and Herzegovina.

The map in Chapter 1, which shows the location of Bosnia and Herzegovina in its regional context, is an important tool for understanding the plight faced by countries occupying the Balkan Peninsula. The map shows Turkey as a land bridge between Asia and Europe. Great civilizations such as the Ottoman Empire, which was centered in Turkey, left their imprint on Bosnia and Herzegovina. So did many other great civilizations and powers; including the Roman Empire, the Byzantine Empire, the Austro-Hungarians, the Germans, and finally, the Soviets.

EARLY HISTORY

Archaeologists have found evidence of people living in the region that dates back some 200,000 years. These people lived during the Paleolithic Age and were hunters and food gatherers. The Illyrian civilization, with origins in the second millennium B.C., was the first to flourish in the territory of modern-day Bosnia and Herzegovina. A thousand years later, Romans called the region, which by then was under their control, Illyricum. These people spoke Indo-European dialects and came from northern Europe. The Illyrians were considered warlike, and often committed acts of piracy on the seas. In 6 A.D., they attempted to revolt against their Roman masters. The revolt failed, but despite falling under Roman rule, the Illyrians retained a high degree of autonomy and independence. After being conquered, many Illyrians employed their warring skills by fighting in the Roman army. Trade routes developed in the region, as goods moved between Eastern and Western Europe. Ports developed on the Balkan coast of the Adriatic Sea, which further enhanced the trade going through and across the peninsula. Contemporary Albanians are descendants of the Illyrians.

The Roman civilization declined in the fourth and fifth centuries A.D., and German invaders, the Visigoths, sacked Rome in 476. The Goths controlled much of the Balkan Peninsula for over a hundred years, until the Slavs arrived in the sixth century. Slavs changed the region forever as they arrived and conquered all of Illyricum. The Slavic culture quickly spread, replacing the previous ways of life practiced by residents of the region. By the ninth century A.D., much of the earlier culture had disappeared in the region, except for a few groups in present-day Albania.

Slavic tribes of Croats, Serbs, and Slovenians were the major players in the region after the sixth century. These tribes moved into the region from the north, and organized themselves into clans that were primarily engaged in farming and agriculture. The Slavs also brought with them their Indo-European language, which is the origin for the language spoken in Bosnia and Herzegovina today.

Starting in 960, a revolving door of rulers moved through Bosnia and Herzegovina, as Serbia, Croatia, the Eastern Roman Empire, Duklja (modern-day Montenegro), and Hungary all governed the region at different times. Finally, after the period of Hungarian rule, Bosnia emerged as an independent country. The country's strength was at its greatest under Ban Stefan Tvrtko, who greatly extended Bosnia's rule over Herzegovina and claimed parts of Croatia and Dalmatia. Tvrtko made agreements with other nations and established a strong medieval

Bosnia during his lifetime. After Tvrtko's death in 1391, the sequence of rulers and dynastic squabbles weakened the kingdom during the first half of the fifteenth century. These conflicts opened the door for Turkish conquest, and ultimately for Bosnia's fall, in 1463.

THE OTTOMAN TURKS AND ISLAM

In 1463, the Ottoman Turks conquered the area now known as Bosnia and Herzegovina. This began 400 years of Turkish rule, which turned the subjugated Bosnia and Herzegovina into a buffer area between the Islamic East and the Christian West. Many people adopted Islam during this era, resulting in many Orthodox and Catholic leaders fleeing the region, in fear of persecution. Travnik and Mostar were established as Turkish capitals, but few Turks actually immigrated to Bosnia and Herzegovina. Some who did established large agricultural estates and began extracting income from the landless peasants who worked the land. Life for most Bosnians worsened under the rule of the Turks. As Turk nobles and religious leaders accumulated power and wealth, the peasants became increasingly poor and powerless.

After nearly 400 years of Turkish rule, a reawakening of nationalist instincts occurred in the Balkans. Montenegro, Serbia, Bulgaria, and Bosnia and Herzegovina rebelled against Turkish rule in 1875–1876. These countries, aided in part by their fellow Slavs, the Russians, defeated the Turks and pushed them from power by 1878. Russia played an important role in this victory, because it desperately sought access to warm-water ports on the Adriatic Sea and not because it was a crusading liberator. During the Congress of Berlin, the meeting that ended the Russo-Turkish War, it was decided that Bosnia and Herzegovina would be governed and occupied by the Habsburg Monarchy. Many European countries wanted to keep Russia out of the Balkans, and also wanted to keep Russian power from expanding.

The Austrian (Habsburg) rule, mandated by the Congress of Berlin, was imposed by force. Benjamin Kallay, a Hungarian, administered the region of Bosnia and Herzegovina until 1903, during much of this period of occupation. He did many positive things for the region; including enforcing laws, and building roads, schools, and railways. Sadly, he also assumed a negative role by playing off the different nationalist groups against one another. He exploited the differences between the Orthodox Serbs, Muslim Slavs, and the Catholic Croats, dividing the people as a means of maintaining Habsburg control over them. Kallay's divisive policies lay the groundwork for ethnic and religious divisions and left a tragic legacy for twentieth-century Bosnia and Herzegovina.

WORLD WAR I

The people of Bosnia and Herzegovina resented outside rule, and detested the Austrian military occupation and rule by the Habsburg monarchy. To make matters worse, much of the Balkan region, including Bosnia, was annexed into the Habsburg Empire in 1908. Serbia had also wanted to annex Bosnia, so the annexation by the Habsburgs enraged the Serbian government and incited further opposition to Habsburg rule in the region.

In the late-nineteenth and early-twentieth centuries, many countries formed alliances with other nations, in efforts to protect themselves and their interests around the world. For example, Germany, the Habsburg monarchy, and Italy formed the Triple Alliance in 1882. Romania later joined this alliance, which was in part a German attempt to isolate France due to a territorial conflict over the Alsace and Lorraine region. In response, France and Russia also formed an alliance, the Entente, because of their fear of Germany, and also because of Russia's dissatisfaction with the Habsburgs' occupation of the Balkans. Like most of the people in the Balkans, the Russians are Slavs, and this kinship strongly influenced Russia's interests in the region. In 1904, the United Kingdom joined France and Russia to form the Triple Entente. The British were becoming isolated with the various alliances being made, and knew that Germany had naval aspirations that could challenge the British, if left unchecked. The Triple Entente and the Triple Alliance were mergers of powerful, yet fearful, countries. They created the alliances to further their interests, which included their own protection. Thus two major "dominoes," the Triple Alliance and the Triple Entente, were in place for the chain reaction of events now called World War I.

On June 28, 1914, Archduke Franz Ferdinand, nephew of Franz Josef (Emperor of Habsburg), was visiting Sarajevo. He was traveling with his wife Sophie, the Duchess of Hohenberg. There had been rumors of a possible assassination attempt on Ferdinand, who was expected to assume the throne after Franz Josef. This was something few Serbs wanted to happen. A secret Serbian society, the Black Hand, had trained a number of young men to enter Bosnia with the assignment of assassinating Archduke Ferdinand. In Sarajevo, a Black Hand member, Nedjelko Cabrinovic, tossed a bomb into the Archduke's car. The bomb deflected off the Archduke and exploded behind the car. Neither Sophie nor Ferdinand was hurt in the attempt. Cabrinovic quickly committed suicide to avoid capture and punishment. However, a little while later the Archduke's driver became lost, and the vehicle turned around to proceed in the proper direction. Surprisingly, another assassin, Gavrilo Princip, happened to be near the car as it was turning around on the bridge. He fired two gunshots. The first killed Sophie and the second her husband, Archduke Franz Ferdinand. Princip was quickly captured, but the first domino in the chain that toppled to start World War I had been pushed over.

Headlines in the *New York Times* on June 29, 1914 screamed, "Heir to Austria's Throne is Slain with His Wife by a Bosnian Youth to Avenge Seizure of His Country." The world was shocked, and the elements were now set for a devastating

In a photograph taken only minutes before his assassination in June 28, 1914, the Archduke of Austria, Franz Ferdinand, walks to a car with his wife Sophie, who was also shot and killed in the attack.

sequence of events involving the Triple Alliance and the Triple Entente. Within a short time, the following major events occurred, throwing much of the world into the chaos of World War I.

Within a week, July 29 to August 4, 1914, the world had gone from relative peace to having seventeen million men and eight countries engaged in "The Great War," as the *New York Times* called it on August 5, 1914. World War I had started, and Sarajevo, in Bosnia, was the city that served as the trigger for the war that lasted from 1914 until 1918. By the end of the war, over 70 countries would have participated and over eight million people would have died.

After the war, Bosnia and Herzegovina became part of a short-lived union of Slovenes, Croats, and Serbs, which several months later, in late 1918, joined the kingdoms of Serbia and Montenegro in a new union called the Kingdom of Serbs, Croats, and Slovenes. The hope was that these groups, all Slavs, could unite in a spirit that was called "Brotherhood and Unity" at the time. At first, this unification made great sense, but underneath lay deep religious and nationalist instincts that undermined their unity. Later, in 1929, the name Yugoslavia was adopted for the country.

Germany invaded Yugoslavia in 1941, in the early stages of World War II. Bosnia and Herzegovina was incorporated into Croatia, which was controlled by the Nazis and the Croatian nationalists called the Ustachi. During the war, large resistance forces, organized by the Communist Party, fought successful guerrilla actions from their bases in Bosnia's remote mountains.

After the war, with Germany and her allies solidly defeated, Bosnia and Herzegovina became one of six republics in a new, Communist-controlled, Yugoslavia. The country was led by Josip Broz, better known as Marshall Tito, who had served as the leader of the Communist Party and of the resistance movement. After the war, Tito emerged as one of the region's strongest leaders. Remarkably, he held together Yugoslavia's different cultural pieces, now divided into six republics, with a continued focus on brotherhood and unity. Unlike many other Communist leaders in Eastern Europe, Tito was able to keep the Soviet Union at arm's length. Part of his success was the result of having buffer countries, such as Romania and Hungary, physically separating Yugoslavia from the Soviet Union. But Tito's strong leadership also was responsible. His autonomy was further exhibited in 1961, when he cofounded the Non-Aligned Movement, along with Egypt's Gamal Abdel Nasser, and India's Jawaharlal Nehru. For many years during the Cold War, his independence put Yugoslavia at the heart of the Non-Aligned Movement, between the Soviet Union and the United States.

Tito's most difficult task was holding together this diverse country. Religious and nationalist differences were great within Yugoslavia's six republics. He was successful in subduing nationalist uprisings when they arose. He also attempted to build unity between and among the country's various antagonistic groups. During his rule, Tito served as the glue that united Yugoslavia's diverse people. However, many people doubted whether Yugoslavia would remain united after Tito's death in 1980.

4

Disintegration and War

Only a short time after the death of President Tito, Yugoslavia started to disintegrate. The glue provided by Tito was gone, and opportunists who had been waiting in the wings started playing upon the religious differences. When a visitor talks to people from Bosnia and Herzegovina today, residents will often say that the later years under Tito, and the first few years after his death, were probably the best that they can remember for Yugoslavia. The economy was stronger than that of other Eastern bloc countries. Additionally, the nation of six republics was well-regarded in many international circles, due to Tito's influence in the Non-Aligned Movement. Quality of life was reasonably good, and the hosting of the 1984 Winter Olympics in Sarajevo put Yugoslavia prominently on the world stage. Tragically, that all ended in the 1990s, with an era of national death, horror, and disintegration.

The opening ceremonies for the XIV Winter Olympics took place in Sarajevo's Kosevo stadium on February 8, 1984. The Winter Olympics put Yugoslavia prominently on the world stage, and it was such a positive experience for the country that Bosnia and Herzegovina hopes to again host the games in Sarajevo in future.

Imagine a simple equilateral triangle representing Bosnia and Herzegovina. Within the triangle, a number of elements held Yugoslavia together. Tito was a strong unifying force, as were the Olympics and the spirit of Slavic brotherhood. Outside of the triangle were forces pulling Yugoslavia apart—religion, regionalism, and nationalism. Differences between the Croats, Serbs, and Bosniaks were deemphasized under Tito's rule. He squelched divisive nationalist activities. But with Tito now dead, others rose to carry the banner of nationalist and religious differences forward, in a manner that resulted in the deaths of hundreds of thousands of people in the 1990s. How did this happen?

SLIPPING TOWARD CONFLICT

In 1989, Slobodan Milosevic was elected president of Serbia. He espoused a new brand of Serb nationalism that intimidated and frightened the Croats, Bosniaks, Slovenes and other minorities in the Balkans. Many of these minorities then formed political parties to advocate for their own groups.

In 1990, Yugoslavia held its first fully free elections. In these elections, nationalist Serb and Croat political parties decisively defeated the Communist Party, the party of Tito. Remember that Yugoslavia was composed at this time of six republics; including Serbia, Montenegro, Croatia, Macedonia, Slovenia, and Bosnia and Herzegovina. In the religiously and ethnically diverse Bosnia and Herzegovina region, a Muslim party was victorious. The Muslims did not have a clear majority in any of the six republics, but represented about 44 percent of the population in Bosnia and Herzegovina. In contrast, other republics in Yugoslavia had one religious or ethnic group that dominated their population. For example, Croats in Croatia and Serbs in Serbia represented strong majorities within these regions.

In 1991, Milosevic began to lay claim to areas with Serb majorities. He started to invade Croatian communities with Serb population majorities. Another horrific chain reaction was started in the Balkans. After seizing these Serb majority communities, Milosevic ordered his forces to take non-Serb majority communities in Croatia, as well. Croatia retaliated in 1992, and soon Serbs, Croats, and Bosniaks were killing each other by the thousands. Atrocities were committed on all sides, in this cataclysmic war that was fueled by nationalist, regional, and religious differences.

INDEPENDENCE AND CONFLICT

In December 1991, Bosnia and Herzegovina officially declared independence from Yugoslavia, which was being led by Serbs, and asked the European Union for political recognition. Bosnian voters supported the declaration of independence with a vote in March 1992, and elected Alija Izetbegovic as president.

Having a Muslim president and Muslim-led government frightened many Serbs in Sarajevo, and on April 6, 1992, Serb militants fired upon a peace rally in the city. Five people were killed, and over 30 were injured in this attack that marked the start of the siege of Sarajevo, a siege that lasted until September 15, 1995. Milosevic's Serbian fighters were relentless in their attacks upon Sarajevo, with "weekend warriors" and other freelance fighters serving as snipers, while the Serb military rained mortar and shells down upon the city.

In peacetime, Sarajevo is a beautiful city with a gorgeous mountain setting in the Miljacka River Valley. The park at the headwaters of the River Bosna is beautiful, with small streams, beautiful vegetation, pools of water, and swans dominating the scenery. During the siege, the mountains sheltered Serbs who fired mortar and guns at the mixed citizenry of the city below. Many Serbs fled Sarajevo on April fifth, having received advance information about the planned attack. Ironically, many Croat and Muslim friends of these people packed lunches and waved farewell to their Serb friends as they left. Non-Serbs did not know that these friends had received an advance warning. This was a tragic parting, as Sarajevo had always been a city of ethnic and religious tolerance. Even today, the citizenry takes pride in their mixed heritage, as there is much intermarriage between Serbs, Croats, Bosniaks, Jews, and others.

The siege of Sarajevo isolated the city from the outside world. All roads were blocked, and the basic necessities of food, water, and clothing were cut off from the population. Sniper's bullets and mass executions killed thousands. Tens of thousands of others were injured, raped, or became victims of starvation during the long siege. The parliament building, Unis Towers, Tito Barracks, and hundreds of other buildings were shelled and bombed, some into dust. The United Nations attempted to intervene by bringing supplies into the airport. But Serb

The park at the headwaters of the River Bosna is one of Bosnia and Herzegovina's many beautiful landmarks, with small streams, lush vegetation, and swans swimming in pools. Unfortunately, after the war it was surrounded by thousands of deadly land mines and their removal could take decades.

fighters kept them at bay for months by controlling the ends of the runways. The Serbs also shot citizens who tried to flee the Miljacka River Valley by running across the airport's runway. This siege was the worst endured by any city in the world since World War II. Even the beautiful park at the River Bosna headwaters was scarred. Thousands of deadly land mines now surround the park, and there is little hope for their removal in the coming decades.

The phrase "ethnic cleansing" was used to describe the effort by one ethnic group to remove another, in order to make an area ethnically homogeneous. This effort was taking place in Bosnia and Herzegovina, and in other areas of the Balkan

Peninsula. People were killed and homes were bombed. When, fearing for their lives, a family abandoned its home, another family of the dominant culture in the region would move in—if, indeed, the house was still standing. This created a huge problem after the war when many displaced persons wanted to return to the homes they had been forced from, only to find someone else living in it.

Fighting spread across the Balkans, including most of Bosnia and Herzegovina. Mostar was devastated and the city's historic bridge was destroyed. Few towns remained untouched by the war. Most communities had at least a house or two that was gutted, in selected instances of "ethnic cleansing" committed against the community's minorities. Finally, in August and September 1995, NATO (North Atlantic Treaty Organization) troops and air forces entered Bosnia and Herzegovina and started bombing Serb military positions. At the same time, Bosniak and Croat military forces reclaimed land in Bosnia and Herzegovina from the Serbs.

THE DAYTON ACCORD ENDS HOSTILITIES

In December 1995, with leadership from U.S. President Clinton, peace talks were held in Dayton, Ohio. These deliberations ended the conflict. They created the Federation of Croats and Muslims and a Serb entity called the Republic of Srpska, within Bosnia and Herzegovina. For a long while, people were afraid to even travel to another entity within the country, as hatred remained high. The situation in the town of Brcko was so contentious that its fate was not even fully resolved by the Dayton Accord. Over sixty thousand NATO military personnel, called IFOR (Implementation Force), were called in to keep the peace and help implement the Dayton Accord. After the peace was secured in 1997, NATO created a Stabilization Force (SFOR), to maintain peace in the region for an unspecified amount of time. Thirty thousand troops were maintained during the early years of the Stabilization Force. Today, people

Serbian President Slobodan Milosevic, Croat President Franjo Tudjman, and Bosnian President Alija Izetbegovic sat together at the ceremony for the signing of the Dayton Accord, which put an end to hostilities in December, 1995.

can move freely between the entities of the Federation and the Republic of Srpska. But much work still must be done before the country's people and regions become united in spirit.

In 1996, Alija Izetbegovic was, once again, elected president of Bosnia and Herzegovina, in an awkward three-person presidency that had been created by the Dayton Accord. Each ethnic group was represented by one of the three presidents. Some Bosnian Serbs resisted the new government. Many resisters, including Radovan Karadzic, former president of the Republic of Srpska, and Serbian president, Slobodan Milosevic, were pursued as war criminals after the war. So were hundreds of others, because of the atrocities they had allegedly committed.

Disintegration and War 49

In this map the Republic of Srpska is primarily the pink areas and the grey and yellow areas are where the Federation of Bosnia and Herzegovina is located.

POSTWAR BUILDING EFFORTS

The aftermath of the war left the young country of Bosnia and Herzegovina in disarray and disrepair. Thousands were dead, the economy was in ruin, and the infrastructure—highways, railroads, power, water, and public buildings—was in shambles. Buildings were scarred with bullet holes, and many

The aftermath of the war left Bosnia and Herzegovina in shambles. Thousands were dead, the economy was in ruins, and much of the country's infrastructure was destroyed beyond repair. Here, a woman puts her laundry out to dry on the balcony of a building that has lost its outer wall.

had been burned or bombed out of existence. Thousands of homes had been gutted. Much of the population had fled. After the war, many of these displaced persons wanted to return home. Nearly one million people had been displaced in Bosnia and Herzegovina, and there were over a million more in other Balkan locations.

Despite the horrors of ethnic cleansing and war during the 1990s, Bosnians tell thousands of stories of kindness. Many of these are simple stories of people helping one another, such as the one told by Sanja, a friend of the author. Her parents were Serb and lived in the entity that today we call the Republic of Srpska. Although Serb, they lived in a neighborhood where

Muslims were the majority. Sanja's parents had always helped their neighbors, in the truest spirit of neighborliness. When the war started, many Muslims left, but one family of Muslim friends lost the head of their family when he died of a heart attack. This left the wife alone, poor, afraid, and in poor health. Sanja's family and others provided assistance to this Muslim woman who was in the wrong place at the wrong time. During the war, Sanja had a young child, and one day, a small gift arrived from the woman who had been helped. She had little to offer because of all her problems, but her gratitude spoke volumes to her Serb friends.

There are many other tales of kindness. Thoughtful acts, across Bosnia and Herzegovina, that helped to save lives, educate children, and provide life's necessities, such as shared food, clothing, and shelter. Thousands of heroic Bosniaks, Serbs, Croats, Jews, and others contributed—often under fear of death—to the well-being of others with different religious or ethnic backgrounds. While war criminals like Milosevic made the headlines, the courage and humanity of these common people can never be forgotten.

The aftermath of the war also left a new flower, the "Sarajevo rose." The war left many blast marks on sidewalks, roads, parking lots, and buildings. After the war, people put a rosy pink plastic substance into the holes in the concrete where mortar rounds had exploded, leaving flowerlike patterns that are now called Sarajevo roses. This distinctive symbol remains today, throughout the city, as a legacy of the war and the tens of thousands of men, women, and children, who suffered and died during the tragic conflict.

Today, there is optimism in Bosnia and Herzegovina. Billions of dollars have been brought into the economy by international organizations. The young people , in particular, have an interest in democracy and in making this society more tolerant of its diversity. With these and other efforts, new flowers are beginning to bloom.

5
People and Culture

Bosnia and Herzegovina's human diversity is the country's greatest strength as well as its greatest weakness. As history has repeatedly demonstrated, this region's cultures can serve to unite the country, as was the case under Marshall Tito. They also can contribute to deadly division and conflict, as evidenced during the 1990s. When united, the country gains prosperity, and it can use its diversity for trade and other economic, political, and social advantages. When the cultures divide, everyone seems to lose in the end, as recent history has clearly shown.

Three different groups dominate Bosnia and Herzegovina's cultural landscape: Serbs, Croats, and Bosniaks. None of the three comprise a majority in the country, and other religious and ethnic groups also are present, but in much smaller numbers. These small minorities, amounting to less than 0.5 percent of the population, include Jews and Gypsies.

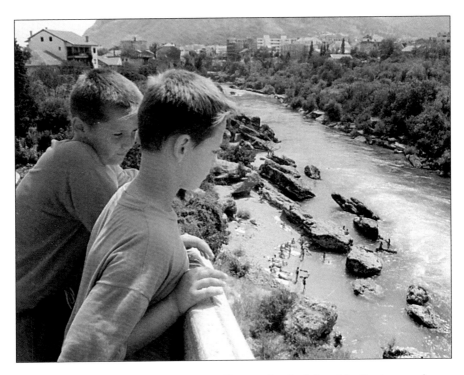

In postwar Bosnia and Herzegovina Nedim Kruskonja, left, a Muslim boy, and his Croat friend Sasa Dukic watch swimmers from a bridge over the river Neretva, which divides the town. Though cultural diversity has caused devastation and conflict, it also can be one of this region's greatest strengths.

POPULATION AND DIVERSITY

Population estimates in Bosnia and Herzegovina vary widely, because of the number of people who fled or were killed during the war. Present estimates, and they are just that, estimates, place the population at about four million. Of this population, 48 percent are Bosniaks, 31 percent are Serbs, and about 14 percent are Croats. These groups are primarily Slavic in cultural heritage, but each has strong connections to a particular religion. Forty percent of the population, primarily the Bosniaks, is Muslim; 31 percent is Orthodox and Serb; 15 percent is Catholic and primarily Croat. Notice the close correlation between the percentages of each ethnic group and their associated religious affiliation. Only 4 percent of the population is Protestant, and the other 10 percent are Jewish, nonreligious, or other religions.

Life expectancy in Bosnia and Herzegovina is actually quite high. The average life expectancy for women is almost 75 years; for men it is 69 years; and the national average is 72 years. These data may be misleading, because they do not include all the statistics from the war years. The long life expectancy is quite surprising in view of the large number of citizens who chain-smoke cigarettes. The country is filled with people smoking in coffee shops, cars, restaurants, businesses, and other locations, in a manner unlike anything existing today in the United States or Canada.

The language spoken in Bosnia and Herzegovina is Serbo-Croatian. However, each ethnic group has tried to take the common language and make it their own. Thus, the Croats speak Croat, the Serbs speak Serb, and the Bosniaks speak Bosnian. In reality, they all sound alike, and are basically the same language. One significant difference is in the written language. Many Serbs use the Cyrillic alphabet much like the Russian alphabet, which is in the same family group. For example, the Serbo-Croatian phrase for "good morning" is *dobro jutro* and is pronounced "do-bro yoo-tro." The Russian word for "good morning" is pronounced "doh-breh oo-tra." There are similarities between many Russian and Serbo-Croatian words, such as the words for numbers, days of the week, and many other terms. This is because all are derived from Slavic linguistic roots. Technically, Bosnians can now claim to speak three languages, but in reality, they speak the same Serbo-Croatian language that was spoken by all before the war. At the same time, they are able to understand many words and phrases from other languages in the South Slavic family. Although Serbo-Croatian is the native language for Bosnia and Herzegovina, many people are bilingual, as they have learned another language, such as English, German, or French.

Each of the ethnic groups has some of its own traditions,

as well as those shared with others. Since a religion is so closely associated with each dominant ethnic group, the two are discussed together.

BOSNIAKS

Turks introduced the religion of Islam to Bosnia and Herzegovina during the fifteenth century. Over the next few centuries, many Christians voluntarily adopted Islam. Among the compelling reasons were tax relief and generally easier living conditions than those experienced by Christians. Many other Christians left the region. In another Turkish practice, tens of thousands of young men went into the military, where they were taught Islam. Eventually, thousands of these youth also converted to the faith of the Turks.

Today, 40 percent of all Bosnians are Bosniaks, or Muslims. The distinction between Bosnian and Bosniak must be clarified. "Bosnian" is the term used to identify any citizen of Bosnia, whereas "Bosniak" refers to Bosnian Muslims. In 1973, the Yugoslav government gave ethnic minority status to people of the Islamic faith. In essence, this act created an ethnic group on the basis of its religion. When civil war broke out in the early 1990s, Muslims in Bosnia were left without a homeland. Croats could claim Croatia, and Serbs could do the same in Serbia. In order to justify their claim to territory, Bosnian Muslims began calling themselves Bosniaks, a final step in their transformation from a religious to an ethnic group. Before the war, many of these Bosniaks were rather tentative in their religious practices and beliefs. Today, many continue to be very moderate in their religious practices. There is no evidence of extremist elements such as those existing in other Islamic areas of the world.

Bosnian Muslims may look very different from the stereotypical Arab face that many associate with Islam. Bosnian Muslims are Slavs and do not differ in appearance from the country's general population. For example, it is very common to see a blond, blue-eyed Muslim in this country.

Today, 40 percent of all Bosnians are Muslims. These worshippers gather at the King Fahd Mosque and cultural center during the opening ceremony and first Friday prayer in Sarajevo on September 15, 2000.

Mornings in Sarajevo, Mostar, and other Bosnian cities in the Federation, greet the visitor with the melodic harmony of the call to prayer. Five times a day, Muslims are called to prayer, in one of the practices taught by Islam's most holy book, the

Koran (Qur'an). Muslims hold Allah as the one God of Islam, and Mohammed is believed to be the prophet of Allah. Muslims believe in these five pillars of Islam:

1. The testimony of faith; Allah is the one true God.

2. Prayer five times a day. Once each at dawn, noon, mid-afternoon, sunset, and night.

3. Giving support to the needy.

4. Fasting from dawn to sunset during the month of Ramadan.

5. Making a pilgrimage to Mecca, Islam's most holy city located in Saudi Arabia, at least once during a person's lifetime.

Ramadan is the ninth month of the Islamic calendar, which is a lunar calendar, and different from the traditional solar calendar used in most of the world. Eid al-Fitr is celebrated as a full-day feast at the end of Ramadan. Another important Muslim celebration is the Hajj. This takes place in the twelfth month when Muslims go to Mecca to circle the Kaaba, Islam's most holy site. Eid al-Adha is the celebration marking the end of Hajj.

Muslims in Bosnia and Herzegovina came under pressure from Serbs and Croats as Turkish rule fell into decline. During World War II, Bosnian Muslims fought on all sides at different times of the war, many being killed by both Communists and Fascists. With the Allied victory over Germany, Tito's rise to power was a positive one for Muslims in Yugoslavia in terms of their safety. However, Tito also opposed all religions, so Islamic schools and Islamic religious manifestations, such as the veil for women, were discontinued. In 1961, however, Tito helped found the international Non-Aligned Movement, at which time he reached out to support Islam in Bosnia, as a way of courting favor with other Islamic countries that were engaged in the

Non-Aligned Movement. In 1968, Tito formally recognized Bosnia as the sixth republic within the country of Yugoslavia.

With the death of Tito in 1980, the religious protections afforded Muslims and other religions gradually gave way to nationalist movements on the part of Croats and Serbs. Slobodan Milosevic pushed Serbs within Yugoslavia into a nationalist fervor in the late 1980s. Many Croats in Yugoslavia quickly followed the Serbs, with their own nationalist moves under Franjo Tudjman. Caught in the Serb and Croat crossfire, Muslims were put on the defensive during the early disintegration of Yugoslavia. Thus, Bosnia and Herzegovina declared its independence late in 1991, and Alija Izetbegovic, a Muslim, was elected president of the new country in 1992.

The complicated birth of the new country and the election of the new president also revealed the deep divisions between the three major ethnic/religious groups. Today, many political parties still reflect the ethnic and religious ties. Muslims now have other political parties such as the Muslim-Bosniak Party and the Liberal Bosniak Party. This entrenchment of religion and politics serves as a continuing divide in the effort to address the common good of all citizens and the country as a whole.

THE CROATS

Croats compose the smallest of Bosnia and Herzegovina's three major ethnic groups, with about 17 percent of the country's population. The Croats were a tribe of Slavs, led by Chief Hrobatos (also spelled Chrovatos), who arrived in the Balkans with the Serbs in the seventh century and displaced or assimilated the former landholders, the Illyrians. King Tomislav, who ruled in the tenth century, created the medieval Croatian Kingdom, which controlled all territories of modern-day Bosnia. At that time the Croats had already accepted religious influences from Rome. The Croatian Kingdom lasted until the end of the eleventh century, when Hungarian rule spread throughout the Croatian territory. During the era of the Ottoman Turks, the Croats maintained control of the coastline and areas now held by the country of Croatia. The present-day border between Bosnia and Herzegovina and Croatia is located near the dividing line that existed between the Croats and Turks during the Ottoman Empire.

World War II found the Croats aligning themselves with the German Nazis and Italian nationalists. This put Croatia on the side of the Axis powers (the three powers of Germany, Italy, and Japan during World War II), which ultimately lost the war. However, it also allowed the Croats to capture the dream of an independent Croatia, for a brief period of time. In truth, this brief period of nationhood was a bit of an illusion because, at the time, Croatia was little more than a puppet government under German control. After the war, in 1946, Croatia became a part of Yugoslavia and was placed under the communist rule of Marshall Tito. When Croatia declared its independence in 1991, tensions heightened between the Serbs and Croats. Independence greatly concerned Serbs living within the area claimed by the new country of Croatia, which led to the armed rebellion against the Croatian government.

Croats in Bosnia and Herzegovina live throughout the Federation, although very few have resided in the Republic of Srpska since the war. Many Croats live in the western region of the country, in Livno and other communities in the Herzegovina region of the Federation. Here, close to the country of Croatia, many Croats were slow to warm to the idea of being a citizen of Bosnia and Herzegovina, in which they were a distinct minority. In the early years after the war, political posters could be seen for Franjo Tudjman, even though he could not run for office in Bosnia and Herzegovina. Croatian money and stamps were even used in Croat regions of Bosnia and Herzegovina, as many Croats passively resisted their new citizenship in the country. Even today, Croats are very likely to cheer on the national soccer team of Croatia over their own country's team, as this cultural connection remains strong.

Catholic Croats celebrate traditional Christian and Catholic holidays, including Easter and Christmas. Their practices are similar to Catholics around the world as they attend weekly Mass, which is conducted by priests.

Like the Bosniaks, the Croats also have their own political parties and are one of the three groups participating in the national three-person presidency. Their parties include the New Croatian Initiative, the Croatian Democratic Community of Bosnia and Herzegovina, and the Croatian Peasants' Party of Bosnia and Herzegovina.

THE SERBS

The Serbs are the other major ethnic group in Bosnia and Herzegovina and make up nearly one-third of the population. Serbs are primarily Orthodox Christians, and they comprise the largest cultural group in the Republic of Srpska. The Republic of Srpska is one of the two major entities, statelike political regions within Bosnia and Herzegovina.

Orthodox Christians believe that their faith teaches the true doctrine of Christ. Thus, an Orthodox Christian believes in the original teachings of Christ. Orthodox Christians celebrate many of the same holidays, as do Catholics and Protestants. However, the dates on which holidays fall are different, because the Serbian Orthodox Church uses the Julian calendar. Christmas is celebrated on January 7, Easter is around April 25–27, and New Year's Day falls on January 14, in the Orthodox tradition.

With the onset of the 1990s war, many Serbs fled for their own safety from Croatia and the Federation of Bosnia and Herzegovina to the Republic of Srpska or to what was left of Yugoslavia in Serbia and Montenegro. At the same time, Croats left Serb-held areas, fleeing to Croatia and western Bosnia and Herzegovina; and Bosniaks left Croatia, the Republic of Srpska,

Orthodox Christianity is the most common religion for Serbs. Here, the head of the Serbian Orthodox Church, Patriarch Pavle, prays during Saturday mass in the main Orthodox Church in downtown Sarajevo, September 14, 2000.

and Serbia and Montenegro, to seek safety in the Federation of Bosnia and Herzegovina. Intimidation, bombings, threats, torture, and even concentration camps forced this mass relocation of people, as they struggled to stay alive.

Like the Croats, the Serbs were Slavs who migrated into the Balkans in the seventh century. The Byzantine Empire exercised a strong influence, and Serbs soon adopted the Orthodox Christian faith of the Empire. In the eleventh century, the Serbs and the Croats parted ways religiously. Serbs embraced Orthodox Christianity and had a separate kingdom, whereas the Croats were Catholic and under Hungarian control. With the arrival of the Ottoman Empire and Islam,

Serbs tried to maintain control. They were eventually defeated, however, in a climactic battle in Kosovo in 1389. With Turks as their masters, Serbs were under foreign domination until the nineteenth century, when they fought for independence. The Congress of Berlin, which ended the Russo-Turkish war in 1878 and placed the Serbs under the Habsburg Empire, stymied their hopes.

Serb nationalism continued to fester until World War I. The situation exploded when Gavrilo Princip, a member of a militant Serb nationalist group named the Black Hand, pulled the trigger that started the chain of events leading to the war. Princip, a Serb, assassinated Archduke Franz Ferdinand of the Hapsburg Empire. After the war, Serbs were incorporated into the Kingdom of Serbs, Croats, and Slovenes. This kingdom lasted from 1918 until 1929, when the country was renamed Yugoslavia. Throughout the history of the kingdom and most of the era of Yugoslavia, the nationalism of Croats and Serbs has brought frequent clashes, both in terms of political conflict and actual violence.

With Serb-dominated Serbia and Montenegro under his rule, Slobodan Milosevic rekindled Serb nationalism in the late 1980s in an attempt to dominate all of Yugoslavia. Instead, fearing Serb domination and nationalism, parts of Yugoslavia began to break away and declare independence. Slovenia and Croatia declared independence in 1991. Macedonia and Bosnia and Herzegovina broke away in 1992. In a move that was both surprising and a source of outrage to Milosevic, many Bosnian Serbs also voted to break away from his Serb-dominated government; they feared what nationalist extremism could do.

Today, in Bosnia and Herzegovina, Serbs are one of the three primary ethnic groups written into the constitution, along with the Croats and Bosniaks. They also have ethnically based political parties, such as the Serb Democratic Party, the Serb Civic Council, and the Serb Radical Party, which advocate various Serb viewpoints and interests.

Other, smaller cultural groups exist in Bosnia and Herzegovina, but Croats, Serbs, and Bosniaks are the major entities. Their division or unity on political, economic, and social questions will serve as the basis for the development of the country. History shows frequent and deep divisions between them. Yet today, many Serbs, Croats, Bosniaks, Jews, and Protestants are working together to develop a multicultural society that respects differences in religion and other aspects of culture. The future of Bosnia and Herzegovina clearly depends on their success.

CHAPTER 6

Government and Politics

Have you ever been forced to do something you didn't want to do? Would you resist, even if you knew that the thing you were resisting was probably in your best interest? This is the dilemma that Bosnia and Herzegovina faces with the new constitution created by the Dayton Accord. This constitution was, essentially, forced upon the country in 1995 as a means of ending the ethnic conflicts and fighting. In this chapter, we will learn about the structure and functions of the government, along with the role of citizens in Bosnia and Herzegovina.

Three-headed animals usually appear only in fictional tales. Perhaps there is a good reason. With three heads and only one body, it might be difficult to determine which of the heads makes decisions. Disagreements between the three heads might eventually tear apart the body of the beast, thereby causing its own destruction. This problem now faces Bosnia and Herzegovina, because the

In January 2002, Bosnian Muslims gathered in front of the Parliament Building in Sarajevo to protest the extradition of six Arabs to the United States on suspicion of terrorist involvement. The Parliament Building still shows war damage years after the end of hostilities.

constitution imposed by the Dayton Accord created a three-headed national government.

After the war, the parliament building in Sarajevo was left with shattered windows although the bombed-out skeleton of the building remained. The government was in a similar state of ruin. The entity called the Republic of Srpska used its own currency and stamps, even though it was, supposedly, a part of Bosnia and Herzegovina. In western parts of the country, in the Croat Herzegovina region,

Bosnian Croats living close to the country of Croatia have found it difficult to warm to the idea of being citizens of Bosnia and Herzegovina. This campaign poster for Franjo Tudjman was hung up in Livno, Bosnia and Herzegovina even though Tudjman was running for office in Croatia and could not run in Bosnia and Herzegovina.

people were using the Croat currency and postage system, even though they lived in Bosnia and Herzegovina. The system had broken down, and various regions were taking governmental matters into their own hands. The nation's cars did not even have a common license plate. Worse, the country could not even agree on the design of its flag. Foreign leaders, including Slobodan Milosevic in Serbia and Franjo Tudjman in Croatia, encouraged the divisions within Bosnia and Herzegovina, in hope of carving out the country's Serb and Croat populations for their own countries.

The country's constitution addresses the challenge of diversity in the preamble by stating the following: "Based on respect for human dignity, liberty, and equality, dedicated to peace, justice, tolerance, and reconciliation, convinced that democratic governmental institutions and fair procedures best produce peaceful relations within a pluralist society, . . . Bosnia[k]s, Croats, and Serbs, as constituent peoples (along with Others), and citizens of Bosnia and Herzegovina hereby determine that the Constitution of Bosnia and Herzegovina is as follows . . ." Thus, the first words of the preamble address the challenges imposed by diversity, which are of paramount importance to the well-being of the people and their government.

The constitution establishes a democratic republic government with two political subdivisions called *entities*. The two entities are the Federation of Bosnia and Herzegovina, and the Republic of Srpska. Sarajevo is designated as the national capital.

THE NATIONAL GOVERNMENT

The constitution of Bosnia and Herzegovina establishes three major components of national government. These include the Parliamentary Assembly, the Presidency, and the Constitutional Court. Each of these branches is somewhat different than those typically found in other countries, because of the challenges faced in the past by religious and nationalist extremism.

The Parliamentary Assembly

The Parliamentary Assembly is composed of two houses, the House of Peoples and the House of Representatives. The House of Peoples is composed of fifteen members, with ten coming from the Federation and five from the Republic of Srpska. The Serbs, Croats, and Bosniaks are each allocated five of the fifteen seats in the House of Peoples. Thus, five Croats and five Bosniaks come from the Federation, and five Serbs come from the Republic of Srpska. Some people question whether this constitutional provision discriminates against Jews,

Gypsies, and others who are not included in the three major groups. They have almost no opportunity to be elected to the House of Peoples. This question may need to be addressed in the future to assure all citizens full political equality.

Members of the House of Peoples are selected from the governments of their respective entities. Thus, the Federation legislative body selects the five Croat and five Bosniak representatives, and Republic of Srpska's National Assembly elects the five Serb representatives.

The process is somewhat different in the House of Representatives, where there are 42 members. Two-thirds (28) of these members come from the Federation, and one-third (14) come from the Republic of Srpska. Citizens from within each entity elect their members to the House of Representatives. In the constitution, the ethnicity or religion of the persons elected is not designated, as it is for the House of the Peoples. Jews and other minorities, therefore, can be elected to the House of Representatives.

The powers of the Parliamentary Assembly include the budget and having the ability to approve or disapprove treaties made by the Presidency. The Assembly can also create laws for the country in a manner similar to that of legislative bodies in other countries.

The Presidency

The Presidency also has an awkward structure that is incorporated by constitutional design. Instead of one president, the Presidency consists of three members. These three must include one Croat and one Bosniak elected from the Federation, and one Serb elected from the Republic of Srpska. The three members of the Presidency are elected for four-year terms, and one member serves as the chair. The Parliamentary Assembly determines how the chair of the Presidency is selected.

The Presidency has the power to conduct foreign relations

and administer laws passed by the Parliamentary Assembly. The Presidency can propose a budget to the Parliamentary Assembly and has the opportunity of speaking to the body when necessary, as do American presidents when presenting the annual State of the Union Address to Congress. The Presidency can also appoint individuals to the Council of Ministers. Ministers head units of the national government and help the Presidency to administer the country's laws. Ministries are much like cabinet departments in the United States and include offices such as the Foreign Ministry, Ministry of Trade, Treasury Minister, and the Ministry of Education.

The Constitutional Court

The Constitutional Court is the highest court in Bosnia and Herzegovina. The Court has nine members. Reflecting concern about the country's troubled past, the court has four members from the Federation, two from Republic of Srpska, and three are appointed by the President of the European Court of Justice with input from the Presidency.

As the highest court in Bosnia and Herzegovina, the justices are empowered with upholding the country's constitution. They are the highest court of appeals and may also exercise original jurisdiction in some cases. This means that there are some cases that they hear for the first time before any other court. Examples of these types of cases are those that result from a conflict between the two entities, or between an entity and the national government.

ENTITY AND LOCAL GOVERNMENTS

The entity governments are much like those of state or provincial governments. They operate at a level below the national constitution, but are responsible for conducting certain powers and duties. Some of these responsibilities are spelled out in the national constitution, such as the requirement for entities to provide a safe and secure environment for people.

The constitution delegates other powers to the entities, including those powers not expressly given to the national government by the constitution. Some of the other duties of entity governments are:

- Establishing special relationships with neighboring countries;
- Providing help to the national government of Bosnia and Herzegovina;
- Providing legal security and protection to people under their jurisdiction;
- Establishing agreements with countries and international associations, with the approval of the Parliamentary Assembly.

In the Federation of Bosnia and Herzegovina, there are local governmental regions, called *cantons*. These operate in the local areas and assume responsibilities that are much like those of city and county governments in the United States. The Republic of Srpska has no similar local unit to the canton, but retains more central decision making in the entity-wide government.

THE RIGHTS OF CITIZENS

The citizens of Bosnia and Herzegovina are guaranteed certain human rights that are stated in the country's constitution. They include:

- The right to life;
- The right to not be tortured or held in slavery;
- The right to a fair hearing in both civil and criminal matters;

- Freedom of thought, conscience, and religion;
- Freedom of expression;
- The right to private and family life, home, and correspondence;
- The right to property;
- The right to liberty of movement and residence;
- The right to education;
- The right to marry and establish a family;
- Freedom of peaceful assembly and freedom of association.

The constitution also expresses that the rights and freedoms listed above are protected "without discrimination on any ground such as sex, race, color, language, religion, political or other opinion, national or social origin, association with a national minority, property, birth or other status." This protection is very important, given the state of the country and the divisions that existed between ethnic and religious groups during and immediately after the war.

A number of international organizations; including NATO, World Bank, Council of Europe, and others have been working in Bosnia and Herzegovina to help ensure that the rights provided in the constitution are protected during the postwar transition period. There are also sections of the constitution that address problems faced by displaced people. These are people who were forced or frightened out of their homes during the war. Others may now be living in their homes, or their homes may have been partially or totally destroyed. The constitution guarantees that displaced persons can return to their homes and reclaim what is rightfully theirs. The owners have the right to be compensated for their losses if the property was damaged or demolished. Even with this provision, the actual return of property has been very slow and is often marked by difficulties.

FOREIGN POLICY

Bosnia and Herzegovina has once again entered the international political scene. The country is a member of the United Nations, World Health Organization, International Monetary Fund, UNESCO (United Nations Educational, Scientific, and Cultural Organization), Interpol, and many other international and regional organizations. Many organizations are also still present in the country, as the international community continues to assist in its reconstruction.

The country has also participated in a number of important international agreements. Among these are the international agreements on air pollution, climate change, hazardous waste, Oceans and the Law of the Sea, Marine Life Conservation, Ozone Layer Protection, and the Nuclear Test Ban Treaty.

Future international possibilities for Bosnia and Herzegovina include potential membership in the European Union. Although a number of steps and criteria must be met before this dream can become a reality, the country would likely derive great benefit if membership were granted.

Think back to the fictional three-headed animal mentioned at the beginning of this chapter. In Bosnia and Herzegovina the three-headed governmental "monster" created by the constitution provides some commonsense answers to the huge ethnic divisions left after the war. Whether the three-headed Presidency and Parliamentary Assembly models used in the country will govern successfully is questionable, but it is possible. If compromise and negotiation between the ethnic groups can evolve, successful governance can follow. If ethnic and religious divisions persist and extend for a long period of time, however, the new constitution may be doomed by ethnic gridlock.

One factor that can assist the country is the development of cross-ethnic political parties. Parties with an eye for the common good of all citizens would greatly strengthen the political process, by decreasing the lines that remain between ethnic and religious groups. The demise and decline of the ethnically tied Croat, Serb, and Bosniak parties, combined with the development of political parties based on political ideology, and a multicultural society, may greatly assist the development of democracy in the country.

The nurturing role of the international community can also help in this governmental transition. Nonetheless, the real power and responsibility remains in the hands of the voters. They can either resent the externally imposed constitution and government, or work with it to slowly improve governmental processes and the ethnic healing process. They can continue to vote for ethnic parties, or instead, cast their ballots for those pursuing the common good of the country and its citizens.

The question of whether the three-headed governmental animal created by the constitution will live in newfound ethnic harmony, or be ripped apart, taking Bosnia and Herzegovina with it, is to be watched closely. Only the power of the voter holds the answer to this question.

CHAPTER 7

Economy

Bosnia and Herzegovina's economy was devastated by the war. Hundreds of thousands of workers and other citizens were killed, and countless others left the country. The transportation and communication systems were in ruin, as bridges and roads had been bombed and railways and communication centers decimated. Factories had been destroyed. The country's currency was in disarray, as the Yugoslav dinar or the Croatian dinar were being used during, and right after the war, in different sections of the country. A strong black market had developed. The government was in disarray and had very little money. Bosnia and Herzegovina faced staggering challenges in rebuilding the economy and the nation.

Fortunately, the international community responded to this postwar trauma. First to arrive were the Implementation Forces (IFOR), that served to create and keep the peace by policing Bosnia

The Bascarsija is in the old Turkish area for shopping in Sarajevo. The area today has many locally made items that are offered for tourists. Older residents will often sit in this area to drink coffee, watch people shop and to feed the pigeons.

and Herzegovina. These forces were followed by the Stabilization Force (SFOR), that provided a long-term external police function. Then, a variety of countries and international organizations began helping Bosnia and Herzegovina to redevelop its government, economy, and social cohesion. Nongovernmental organizations (NGOs) played a key role in helping the country to rebuild and heal. NGOs operate outside of government and include such agencies as the Red Cross, religious organizations, and many other groups that provide relief in the aftermath of war.

One of the most pleasant and interesting elements of the economy is the local market. Here, a person can buy all types of consumer goods and food products. Clothing, flowers, meat, breads, cigarettes, shoes, toys, household goods, and wonderful fresh fruits and vegetables are available. A walk through the market provides the visitor with sights and smells that capture the ambiance of the marketplace. The market is also a social environment, where friends meet and talk. One or more markets serve most communities. Towns and cities also have locally owned cafes, coffee shops, and stores that offer a variety of goods and services. Most international chains, such as Kentucky Fried Chicken and McDonald's, do not operate in the country at this time because of the war and political and economic uncertainty.

Most Bosnians cook at home instead of eating at restaurants, but many good local cafes are opening as family businesses. Common foods include Bosnian pizza, Bosnian pot (a stew), and other delights that feature various meats and vegetables. A great local meal is the mixed grill, that serves up chicken, veal, pork, beef, and fried potatoes. The chopska salad features fresh vegetables with a local cheese scattered over the top. Visitors enjoy Bosnian foods and, like the locals, frequent the restaurants that specialize in local dishes.

Income levels in Bosnia and Herzegovina are low by international standards, as a great percentage of the population remains unemployed. Data on the average income, unemployment, and other economic data are not readily available in the postwar era. Vital information collection networks are not yet in place. Other problems exist, as well. For example, data are collected separately by the two entities, the Federation and the Republic of Srpska. This makes much of the national economic information unreliable.

Many citizens, including the youth and unemployed, are often discouraged by the economy's slow improvement. Countless factors must yet be addressed if the international community is going to begin investing in the rich potential offered by the country's rich human resources. This chapter briefly reviews some of the major economic challenges and opportunities that exist for Bosnia and Herzegovina.

AGRICULTURE

Agriculture continues to be a major economic activity in Bosnia and Herzegovina. All types of fruits and vegetables are raised, along with cattle, pigs, and poultry. During the war, and with the loss of meaningful currency, many people depended upon crops and animals for barter. They would trade with each other for products and services that they could not provide for themselves.

Most farms are very small and family run. There is little heavy equipment, because it is so expensive. Horses or other draft animals still can be seen in the countryside, pulling plows, other equipment, and carts. Crops include corn, wheat, potatoes, and a variety of fruits; including apples, grapes, pears, tomatoes, and plums. A number of vegetables also are raised as both subsistence and commercial crops. Much of rural Bosnia and Herzegovina still functions on a near-subsistence or folk economy level. This means that people are largely self-sufficient economically; their chief concern is to meet their own needs. Cattle, pigs, and chickens are raised, as meat is an important dietary staple, which is often served grilled. Wonderful local cheeses are produced, for which many communities have earned a great reputation throughout the country. In the summer, tomatoes are grown and sold in local markets.

MANUFACTURING

Much of the manufacturing that existed in the former Yugoslavia has disappeared because of the war. This has resulted in very high unemployment in parts of Bosnia and Herzegovina, in some places in excess of 80 percent. In many areas, four out of five people are unemployed and seeking work. However, a ripe opportunity exists for foreign investors who want to establish new factories, and desire an eager and talented work force.

Today, there are limited industries in steel, mining, chemical refining, fertilizer, and various cottage industries, including woodworking and textiles. Many handmade products are available in the local Stari Grad, or old town section, of Sarajevo. Items such as textiles, custom-made shoes, carpets, chess sets, and other handmade products are inexpensive for visitors and usually of good quality. Even old shell casings from the war have been turned into pieces of artwork, with carvings etched onto the surface by local artists.

NATURAL RESOURCES

Mining is important to the economy and also serves to create employment in related manufacturing areas. Mineral resources include coal, iron ore, manganese, zinc, copper, lead, and bauxite. Steel is manufactured in the country, and this production has now turned away from producing weapons of war, such as tanks. Instead, consumer goods are now being manufactured from the country's natural resources.

The timber industry is also an integral part of the economy. Most lumber comes from the upland areas, where forests are abundant. Some tree species also bear fruit or nuts, providing another food source for the country.

Energy production is drawn almost totally from the natural environment and resources. Bosnia and Herzegovina is blessed with mountainous and hilly terrain and fast-running streams. This combination is ideal for the production of hydroelectric power, which, in fact, accounts for nearly two-thirds of the country's energy. Fossil fuels account for just over one-third of the country's energy supply.

TRANSPORTATION

Transportation systems in Bosnia and Herzegovina are slowly coming back into service, after the devastation caused by the war. Bridges, roads, train tracks, airports, and transportation buildings were damaged or destroyed during the conflict. Trains within Sarajevo, and those from the capital to cities throughout the country, were out of service for the first few years after the war. Now, many are moving again, especially within Sarajevo and those linking major cities. Taxicabs are easily available in Sarajevo and other larger cities, and the cost is very reasonable. Buses connect many of the cities and towns, and today, an increasing number of cars are seen on the roads. Automobile travel between parts of the country can be slow, but rural travel is worthwhile for the visitor. The terrain is beautiful, offering an everchanging landscape of hills, mountains, rivers, forests, and cultural features. However, visitors may not want to wander too far from the road in those parts of the country where land mines remain a major problem. Most of these areas are well-marked with signs and yellow tape, but every year, people are killed or maimed by unmarked mines.

Air transportation is still limited in Bosnia and Herzegovina. Few citizens have enough money to fly. International flights are available and connect Sarajevo to Hungary, Austria, Switzerland, Turkey, Germany, France, Croatia, and Slovenia. Recent renovations to the Sarajevo Airport have made it a dependable facility.

Inland waterways also serve as important transportation routes. The Sava River, located in the Republic of Srpska in the north, has a number of important port cities: Brcko, Bosanski Samac, Bosanski Brod, Bosanska Gradiska, and Orasje. Many of these waterways are not in good repair, with silt in the rivers and bridges destroyed during the war still posing obstacles to navigation. However, improving the inland waterways for trade holds economic potential for the country.

Improved transportation systems are vital for Bosnia and Herzegovina's participation in foreign trade and for internal movement of goods and products. The terrain and present road system make travel slow. New road and rail development will be expensive and will require international monetary assistance. Yet the physical environment holds great potential for tourist development, as the countryside is a traveler's delight. New roads and railways are essential for development of this potential.

COMMUNICATIONS

Communication is important for all countries. It is a necessity for economic development and essential to developing a cohesive and well-informed society. Like other aspects of the economy, key areas of communication were knocked out by the war. In some respects, rebuilding has been easier, because of satellite technology that requires less capital and infrastructure. A prime example is the, cell, or cellular telephone. Before this technology, societies needed telephone centers with switching technology, as well as a network of telephone wires. Today, a citizen in Bosnia and Herzegovina, like so many others in the world, simply dials a party on his or her cell phone, without the expensive ground lines and other costs of traditional telephone service.

In mass communication, Bosnia and Herzegovina has over thirty television stations and about twenty-five radio stations. Computer access is limited; the Internet and e-mail are generally accessible only though coffeehouses, hotels, and other small businesses that offer these services. Only about 1 percent of the population used the Internet in 2003, but the number is increasing.

In Bosnia and Herzegovina, no newspaper is more famous than Sarajevo's *Oslobodjenje*. This paper was famous for its fairness to all ethnic and religious groups during the war. The newspaper advocated religious and ethnic harmony. During the conflict, Serb nationalists tried to bomb the building that housed it—and the paper—out of existence. But it continued to publish from the demolished site, with Serbs, Bosniaks, and Croats bravely working together to publish the imperiled daily. Today, the building stands as a *de facto* memorial to the courage of the people who advocated the brotherhood and unity that stand as Sarajevo's legacy. The story of this newspaper is captured in the book, *Sarajevo Daily*, eloquently written by Tom Gjelten.

The postal system has improved greatly since the war, and stamp collectors are discovering the beauty of the new stamps issued from Bosnia and Herzegovina. Uniform stamps are now used across the country. This is a huge improvement from the early years after the war, when the Republic of Srpska issued its own stamps, and Croats in Herzegovina, in the west, honored Croatia's stamps.

TRADE

Trade is developing slowly. The country has had to import many goods and products since the war. Following the war's end, there was a huge influx of international assistance, but most of it was in the form of charity, rather than trade. The country was not even self-sufficient in food production after the war, so others provided food products.

Today, exports are still limited to handcrafted products, or other simple manufactured goods, such as chemicals and fertilizer. Imports include food, machinery, industrial products, transportation products, and communications technology. The country has a trade deficit because of the war and its economically devastating effects. Trade relationships are now slowly developing with other countries; including Croatia, Slovenia, Germany, Italy, and Switzerland.

Time, international assistance, a stable monetary system, and foreign investment are some of the potential healers of Bosnia and Herzegovina's economy. International investors want a secure political and economic environment before they provide capital resources for developing factories and other businesses. International aid and assistance has been strong from Europe and the United States. But this aid is easily diverted elsewhere, as these countries, and the world, turn their attention and assistance to new areas, such as Afghanistan and Iraq. The internal ethnic healing is a very slow process, but is essential to further economic development. All of these elements are moving forward and allowing Bosnia and Herzegovina to slowly gain new economic opportunities.

One example of progress is the country's currency. In 1991, before the war, inflation was running at 80 percent per month. This presents an intolerable situation that is detrimental to banking and a developing economy. After the war, the international community imposed the convertible marka as the country's currency. It was based on the German mark, a strong European currency at the end of the war. Even though the German mark no longer exists (Germany now uses the euro), Bosnia and Herzegovina continues to use the convertible marka as its national currency. This monetary system is now based on the value of the euro, but it has provided a stable currency with very low inflation. This factor is very important for the development of a strong economy.

The economy is improving, sometimes at a snail's pace. But a repeat visitor cannot help but notice the positive changes that have occurred. New banks and buildings are evident in Sarajevo and Banja Luka. New businesses, even world-class shops, are slowly appearing. The airport in Sarajevo has a new, bright face, and the pace of daily business and local shops has increased. Transportation and communication systems are improving, including the widespread use of new technology. As political and social conditions continue to improve, local and international entrepreneurs emerge; ready to invest their ideals and capital. Hope is being renewed as the nation pulls itself forward from the ashes of the war. The country's people—its greatest resource—can take credit for their perseverance, patience, and optimism.

Cities

Cities in Bosnia and Herzegovina distinctly reflect the history and culture of the country within which they exist. None of the country's cities can compete with major urban centers in the world regarding population, economy, communications, or trade, yet each has an important story that is worth exploring. Each city captivates the visitor with its own rich environment of culture, history, and physical beauty. Each city has a distinctive character, unique from the others. And each city has a story to tell.

The country's largest urban center is Sarajevo, with 434,000 people (Year 2002 estimate). Banja Luka is second in size, with 250,000, and Mostar is third, with 65,000 (estimated). Other cities worth mentioning include Brcko and Travnik.

BANJA LUKA

Banja Luka is the largest city in the entity of Republic of Srpska,

Banja Luka is the largest city in the Republic of Srpska. The city's history dates back 2,000 years and there are remains of Roman forts and baths evident near the city. Here, in July 2003, a group of Bosniaks gather in front of a mosque that is the first to be reconstructed after the 1992–1995 war.

with over 250,000 residents. The city, located in northern Bosnia and Herzegovina, serves as a major cultural, economic, and educational center for the region. Banja Luka is located in the beautiful Vrbas

River Valley, with both the Vrbas and Vrbanja Rivers running through it.

The city's history dates back two thousand years, as indicated by remains from an Illyrian tribe that were found in the area, dating to 9 A.D. Romans also made their mark on the early history of Banja Luka. Remains of the Roman Fort Castra still exist on the Vrbas River, as a part of the famous local fortification, the Kastel. Remnants of Roman baths are also located near the city. Slavs arrived in the sixth century, adding their cultural flavor. Ottoman Turks arrived in 1528 and held the city for 350 years.

Banja Luka is the regional railway hub and highway crossroads. Besides being a transportation center, Banja Luka engages in other economic activities; including production of textiles, beer, dairy products, tobacco products, and hydroelectric power. Coal is mined nearby. The city is also home to the University of Banja Luka.

In 1993, Serb nationalists destroyed 16 mosques. The most famous was the Ferhadija Mosque, dating to 1580. Many refugees who fled from the Federation or from Croatia during the war remain in the city. Today, the Office of the High Representative (OHR) has a regional office here. Its task is to work with the local population to implement the Dayton Accord in the city and Republic of Srpska. The OHR is a civilian organization, created in 1995 by the Dayton agreement to carry forward the implementation of the accord after the war. Offices are located in a number of cities; including Sarajevo, Mostar, Brcko, and Tuzla.

MOSTAR

Mostar is located in the Neretva River Valley in the south of Bosnia and Herzegovina, in the Federation entity. Mostar has always been a divided city, dominated by Christian Croats and Muslim Bosniaks. The city's name comes from the famous bridge that was built there in the 1500s by the Ottoman

Turks. It was called *mostari,* or *Stari Most,* meaning "city of bridges."

Before the war in the early 1990s, Mostar had a population of over 125,000. Today, the number of residents is about half that of the prewar population. During the war, there was devastating fighting between the Croats and Bosniaks. Hundreds of thousands of rounds of ammunition were fired during the peak of hostilities. Buildings at the front line are still laced with bullet holes and other damage from the intense fighting. Large graveyards now rest just a few blocks behind where the front lines of the fighting took place.

In 1993, Croats demolished the old bridge of Mostar, which for centuries had been a symbol of unity for the city. Today, an Italian corporation is rebuilding the Stari Most bridge, but the river continues to divide the city into two sections—Christian Croat and Muslim Bosniak. The population was much more mixed before the war. That era has passed, however, and only recently, and very gradually, has any intermingling begun to occur between the two groups. It is ironic that in 2003, Mostar, the city of bridges, has no bridges. The international community has had a strong military presence in the city and is working to construct both physical and human bridges.

BRCKO

Brcko is an unusual city. Located on the mighty Sava River, it is presently outside of both the Federation of Bosnia and Herzegovina and the Republic of Srpska, but it belongs to both entities. This placement was designed by the Dayton agreement because of the strategic location of Brcko, at the narrow corridor adjoining the eastern and western sections of the Republic of Srpska. Neither the Federation nor the Republic of Srpska was anxious to relinquish its control over this city. The Federation needed river access for trade with Europe. Republic of Srpska feared losing the city, as it was the connection point between the eastern and western portions of the entity. Hence, the Dayton

In 1993, Croats demolished the old bridge of Mostar, which for centuries had been a symbol of unity for the city. The arched bridge was built by the Ottoman Turks in 1566. Today, an Italian corporation is rebuilding the Stari Most bridge.

Accord put it under U.N. administrative control and not within either of the two entities. Technically, Brcko belongs to both of the entities, as the city is the place where both overlap. This presents an unusual political situation for any community. The city is mandated to be multiethnic, but it is unified under the authority of one local government.

Before the war, the population of the Brcko District was about 90,000, with a mixed community of 45 percent Muslim or Bosniak, 21 percent Serb, 25 percent Croat, and 9 percent other. Some reports have the population dropping to 30,000 by 1994, with Serbs as the dominant group with 93 percent, 5 percent

Muslim, and 2 percent Croat. As is obvious from these changed numbers, the war radically changed the composition of the community. The three ethnic groups continue to be at odds with one another, as evidenced by the city's segregated schools.

Brcko is Bosnia and Herzegovina's largest river port and serves as a regional center for food production and agricultural processing. Corn, wheat, dairy, and livestock are major agricultural products in the region. Brcko was a major trade center within Communist Yugoslavia, but trade declined drastically during and after the war. Currently, efforts are underway to redevelop the city's trade function on the Sava River port.

TRAVNIK

Travnik served as the capital for the Turkish rulers during the eighteenth and nineteenth centuries. The city is located in the center of Bosnia and Herzegovina, and only some 50 miles (80 kilometers) from Sarajevo. Travnik lies in the beautiful valley of the Bosna River and once was home to many attractive churches and mosques. Like so many other places, however, the city was severely damaged by Serb shelling early in the war and in later fighting between the Croats and Bosniaks. Most Catholic and Orthodox churches and Muslim mosques were damaged or destroyed during the war.

The composition of Travnik's population was also greatly affected by the war. The city is an example of the poisonous effects of "ethnic cleansing." Before the war, in 1991, about 70,000 people lived in the Travnik district. Forty-five percent of these were Bosniak or Muslim, 37 percent were Catholic or Croat, 11 percent were Serb or Orthodox, and 7 percent were of another ethnic or religious designation. After the war, Travnik's population dropped to 52,000 people with nearly 79 percent Bosniak, 19 percent Croat, 1 percent Serb, and 1 percent other. Also, in 1999, there were over 13,500 displaced persons in Travnik. These were mostly Bosniaks who had fled their homes in areas captured by Serb or Croat forces. The human tragedies in Travnik number in the thousands, a figure that represents only a small percentage of the actual number of people affected by the war. Today, the city is trying to restore its dignity and prewar complexion as a regional economic center.

SARAJEVO

Sarajevo is the city that clearly reflects the heart and soul of Bosnia and Herzegovina. It holds the best and worst of the nation's history and is a capsule of the country's diversity, complexity, and opportunities. Sarajevo is located in east central Bosnia and Herzegovina, in the heart of the beautiful Miljacka River Valley. The city serves as the capital of the country. The original name of the city was "Hodidjed." "Sarajevo" first appears in mid-fifteenth-century Turkish documents.

It is estimated that the city's population in 1991 was 525,980. By 2003, the estimate had dropped to 434,000. In reality, it is difficult, if not impossible, to determine an actual figure because of the high number of people displaced or killed in the war.

Key points in Sarajevo's history include the assassination of Archduke Franz Ferdinand, heir to the throne of the Austro-Hungarian Empire, and his wife Sophie, the Duchess of Hohenberg in 1914. This event triggered World War I. During the brutal siege of Sarajevo, which lasted from 1992 until 1995, tens of thousands of people were killed or wounded.

One of Sarajevo's historical high points occurred in 1984, when the city served as host for the fourteenth Winter Olympic Games. The two-week event was conducted from February 8–19, and it went off without any major problems. The city proudly and rightfully claims that it hosted the most successful Winter Olympics in the history of the event. Forty-nine nations participated, with over 1,500 athletes engaged in the competition in six different sports with 39 events. The Olympic mascot was Vucko, the wolf. His image appeared throughout the city during this successful event.

Sarajevo is the country's capital and largest urban area, and is located in east central Bosnia and Herzegovina. The original name of the city was "Hodidjed" but it has been known as "Sarajevo" since about the mid-fifteenth century. Though Sarajevo was one of the hardest hit areas during the war, its energetic and courageous people are now leading it into a period of artistic and cultural renaissance.

In an effort to regain the unity associated with the Winter Olympics, during recent years Sarajevo has endeavored to host the event again. The city was unsuccessful in its bid for the 2010 winter games, but may bid again in the future. Some serious problems do remain; for example, the bobsled run is still seeded with land mines! However, because of the city's 1984 success, the Winter Olympic site selection committee may look favorably upon Sarajevo in the near future.

In 1999, United Nations Secretary-General Kofi Annan visited Sarajevo and symbolically credited the city with the birth of Earth's six-billionth person. Then President Bill Clinton visited, as did Princess Diana, the singing group U2, and

U.S. Secretary of State, Colin Powell. Fortunately, the city is recapturing its cosmopolitan prewar traditions and status on the world stage. Art shows and music concerts featuring world-famous artists occur frequently. In August, the city now annually hosts the internationally recognized Sarajevo Film Festival. New buildings are appearing, and old ones are being repaired or replaced.

City transportation systems, including buses and light rail, are working again, and buses connect the city with others throughout the country. The airport has been rebuilt in much the same design as it had during the 1984 Winter Olympics. As the air hub of the country, Sarajevo has international flights connecting to Istanbul, Budapest, Munich, Zurich, Ljubljana, Vienna, and other major centers.

The presence of many international non-governmental organizations (NGOs), the infusion of international aid, and the presence of SFOR forces in the area have bolstered the economy of Sarajevo. The economy also gets a major boost from being the country's capital. Shopping in the city can be a delightful experience. One may stroll through an open city market, enjoying the sounds, smells, and sights of the local economy that includes a variety of foods and other daily necessities. The main shopping district of the city is regaining many world-famous name-brand stores. A visit to the Turkish market called Bascarsija, or Stari Grad, is a must for any traveler, as many treasures from the past and present are available at reasonable prices.

The people of Sarajevo can be credited for the city's energetic rebirth. People who remained during the war and survived the siege have demonstrated courage and perseverance under the most difficult conditions. Today, these people are leading Sarajevo into a renaissance of the arts and culture. Their aspiration is to again host the Winter Olympics and to have the world see the city as being much more than was indicated by the tragic news of the 1990s. The eternal flame, commemorating World

War II, is representative of the city's soul. It is located near the places of worship for four different religions; including Judaism, Orthodoxy, Catholicism, and Islam. The city remains a powerful symbol of how people of varied backgrounds can live in harmony in the same community and enjoy cultural diversity.

Large urban cities and small rural villages, alike, were caught in the ravages of war. It made little difference whether one was Serb, Croat, or Bosniak—all were caught in the crossfire. Many individuals, from each group, committed unspeakable atrocities, but today, they are being held to international and domestic standards of justice. Numerous communities, such as Travnik and Banja Luka, are no longer truly integrated. Others, such as Sarajevo, Mostar, and Brcko, are struggling to develop multi-ethnic communities. Displaced persons are found everywhere in the country, and many are still abroad. Yet the twenty-first century is providing a new opportunity for rebirth.

Each of the cities in this chapter had citizens who were heroes during the war. There were people who protected and helped those who were persecuted. Many left their doors unlocked at night, in case someone needed safety. There were teachers who taught without pay, by candlelight, during the war years. There were principals who crawled to school in their suits, to avoid being shot on the way. Heroes and heroines abounded, during and after the war. When you meet survivors, ask them to tell you stories about the bravery and courage exhibited during the war. Many such tales of heroism and compassion are told today. Every city and village in Bosnia and Herzegovina has these powerful stories to share.

CHAPTER 9

Bosnia and Herzegovina Look Ahead

In examining the past and looking ahead to Bosnia and Herzegovina's future, it may be useful to compare the country to music. In the past—both distant and recent—there has been discord. Many conflicting sounds have created musical chaos. The instruments have had different harmonies and rhythms, factors that have created disharmony and lyrical dissonance on many occasions. The fighting in the 1990s was a continuation of this disharmony. Bosnia and Herzegovina was in chaos; the music was sad and ethnically dissonant, as hundreds of thousands were killed during this war-torn concert.

Your reading has revealed that Bosnia and Herzegovina is a country historically beset by deep, long-standing ethnic and religious conflicts. Today, the country is fragmented. Repairing the shattered pieces of society presents a formidable challenge for the government and citizens of the country. The country's future rests

This group of Bosnian refugee children were part of a group of twenty from Bosnia, Croatia, and Yugoslavia who visited the eastern Bosnian town of Srebrenica in November, 2000 to commemorate the signing of the Dayton Accord five years before.

upon the success or failure of reassembling the various ethnic fragments into a cohesive whole.

From thousands of destroyed buildings, to the Sarajevo rose, to the hearts and souls of people damaged by the scourge of fighting and ethnic cleansing—much work must be done. Although slow in its implementation, international intervention in the former Yugoslavia has brought peace and greater stability to Bosnia and Herzegovina (and elsewhere within the region). With strong American support, the international IFOR and SFOR forces have had remarkable success in bringing peace to the region. Peace and stabilization are, however, only the first steps, as much more work is necessary to rebuild this society.

Major challenges remain for Bosnia and Herzegovina. They include economic development, establishing ethnic and religious tolerance, and further growth of an effective democratic political system. Also, it is essential that an informed citizenry develop, one that is concerned with both individual rights and the common good. These, and other issues, must be addressed in order for the nation to move forward in a healthy and positive manner.

The concert played in Bosnia and Herzegovina's future will be orchestrated on a stage that whispers through the forests and hills of the north and across the agricultural lands in the south. The concert conducted within the country during the 1990s was one of great discord. Yet, in earlier years, as a political component within the greater Yugoslavia, cultures and religions coexisted more harmoniously.

Bosnia and Herzegovina provided the clash of cymbals that triggered World War I. Yet, the country sang a beautiful ballad while hosting the 1984 Winter Olympics. Then came the war and ethnic cleansing of the 1990s, and both its music and lyrics were marked by dissonance. Following the war, others have played much of the music emanating from Bosnia and Herzegovina. The sounds of the future have yet to be determined.

The pace of economic reform in Bosnia and Herzegovina will be slow. The outside world still holds war-torn images of the country. Damaged roads, railroads, and communication systems are slow to be repaired, as money is in short supply. Foreign countries are assisting in these efforts, but as Bosnia and Herzegovina falls off of the radar screen for international assistance (due to higher relief priorities in other locations, such as the Middle East), this aid may decrease sharply. New roads are needed, and the rail system must be rebuilt. With no active ocean port on its coast, trade is limited. Yet, creative minds may find some economic potential for the country's short 13-mile (20-kilometer) coastline on the Adriatic Sea. Communications are improving with the use of cell phones.

However, the cost of a cell phone is beyond the reach of most Bosnians. Internet access is also very limited, and few people are able to afford this luxury. The slow, but low-cost, mail system is still the primary means of basic communication.

Farming and agriculture will continue to be vital economic activities, but the potential for manufacturing is great. With high unemployment and a relatively strong traditional education, Bosnians can provide a strong labor base for venture capitalists and entrepreneurs. Incoming businesses can also benefit Bosnians by helping to create a multicultural society. This effort could be assisted by building internationally funded businesses only in communities with integrated work places, in which Croats, Serbs, Bosniaks, and others are locally available for work. By not rewarding ethnically cleansed, or "pure," communities, the outside investors would be furthering the goal of a diverse society. This means that to get a good job a Bosnian, for example, would need to get along with others from different religious and cultural backgrounds.

Small businesses also continue to develop in Bosnia and Herzegovina. Many people are embarking upon economic ventures started with little cash and a lot of hard work. In this manner, small coffeehouses, bed and breakfasts, and neighborhood stores have opened in many communities. This grassroots economic rebirth can spawn local business successes that reinvest in their own businesses, and can then create other businesses. All of these small ventures and spin-offs will create jobs, and stimulate competition and economic development.

The social backbone of Bosnia and Herzegovina was nearly crushed by the ethnic cleansing and fighting during the 1990s. The differing religious groups became almost foreign to each other. Leaders of Croats, Serbs, and Bosniaks had cruelly constructed towering walls of nationalism. Thousands were crushed and killed by the hatred spawned by these leaders. The remaining animosity presents a formidable barrier in the quest to develop a tolerant, multicultural society. Restructuring a society after these terrible events will not be easy, as many parents still pass ethnic hatred on to their children. Many others see the wisdom of ethnic harmony and pass along these values. The scars are as deep and the walls are high with some individuals. Gradually, however, the extremes of hatred and human conflict are leveling off.

A notable activity is now taking place in Bosnia and Herzegovina, with the introduction of Project Citizen and other civic education programs for young people. Starting in 1996, the program, developed and sponsored by the Center for Civic Education, in Calabasas, California, started training teachers from all three major religious groups. These teachers were trained in integrated religious groups. After the trainings, educators began teaching Project Citizen to their students. In 1997, a national event drew hundreds of Croat, Serb, and Bosniak students to Sarajevo. There, they shared public policies addressing local problems and needs, that they had formulated and recommended.

Initially, students from different religions were suspicious of one another, as stories from the war and religious stereotypes kept a silent wall between them. Soon, however, students began to communicate with one another, the walls began to crumble, and stereotypes began to vanish. Today, the program is called Civitas Bosnia and Herzegovina and tens of thousands of Croat, Serb, and Bosniak students are involved in Project Citizen each year. Since 2001, all tenth-grade students are required to take a new civics course. This course was developed by a mixed group of talented Bosnian educators, with the leadership of Civitas Bosnia and Herzegovina and assistance from the Center for Civic Education. In other programs, younger students are also learning basic democratic ideas.

The work in Project Citizen and civics is not only helping to decrease social barriers between youth of different religions, but it is also helping to develop a national identity for Bosnia and Herzegovina. Youth are developing the knowledge, skills, and attitudes essential to the establishment of democratic

Bosnia and Herzegovina Look Ahead

These high-school students are taking part in a civic education program that encourages students from different religions to communicate with one another and break down religious stereotypes. Today, the program is called Civitas Bosnia and tens of thousands of Croat, Serb, and Bosnian students are involved in the program.

traditions in the country. Ideas such as justice, freedom, participation, and responsibility are providing youth with important democratic cornerstones that will support the growth of democracy and a sense of national identity. Project Citizen and the other efforts in democracy education are providing the country's youth with common civic experiences. This background, it is hoped, will enhance the development of a civic identity inclusive of all ethnic groups.

THE ROAD AHEAD

When examining a country's future, it must be considered that there are unlimited possibilities, hinging upon both predictable and unpredictable events and personalities. Here, two possible futures are considered for Bosnia and Herzegovina.

Clearly, it is possible that Bosnia and Herzegovina could split apart along ethnic or religious lines. This is a process that started with the cataclysmic division of Yugoslavia and the beginning of ethnic cleansing. During the war, many displaced persons sought refuge in areas where their religion or ethnic group was dominant. This division could be started again, with tragic results.

Others believe that the future of Bosnia and Herzegovina will consist of a more integrated and tolerant society. Many international agencies, community leaders, educators, and political leaders are working diligently towards achieving this prized goal. The lessons learned during the 1990s were hard, and few people want to revisit the horrors of that era. When asked about the best years of their lives, many people cite the later years under Tito, or the time following his death when greater ethnic unity and tolerance existed. Under this scenario, the fragment of the former Yugoslavia, called Bosnia and Herzegovina, must learn to live together again, or it will die.

The city of Sarajevo is still an expression of ethnic diversity. It demonstrates clearly how members of a diverse community can coexist. With mixed communities and further intermarriage, the religious lines can blur, and may, someday, cease to act as a divisive element.

Obviously, the second of these future scenarios would be most desirable. Gaining the desired future will require time and hard work in the political, social, and economic arenas. At some point, when integrated political parties develop and strengthen, the ethnic parties will weaken. This can then lead to constitutional reform that will eliminate the need for the ethnic group structures built into the legislative and executive branches.

The return of displaced persons to the homes they occupied before the war will potentially reintegrate communities. This task is difficult and complex, however, as trust has evaporated for many people, and they are reluctant to return to their homes, even with the protections provided by the international community. Extensive work will be needed to bring these communities back to their prewar levels of religious acceptance and integration.

Many of the elements discussed as possible situations for the future are starting to better harmonize. Hopefully, a new concert is being written. The country is dreaming in a positive way again. This is clearly demonstrated by Sarajevo's bid to host the 2010 Winter Olympics. Although the bid was not accepted, it demonstrates that the country is looking forward again with new vision and renewed hope.

A visitor to the country feels this new spirit when taking a korzo (stroll) down the main streets of Sarajevo. There is renewed energy on the streets. People are dreaming again and beginning to take those forward steps that will help themselves and their families.

One of the most memorable evenings the author experienced in Bosnia and Herzegovina was in a café around the Orthodox New Year. In a group of educators, consisting of people from diverse backgrounds, a Serb and a Muslim were playing guitars, and people of all religions were singing together. Voices blended together in harmony, in songs of love, country, and tradition. The voices of loving Bosnians named Zdravko, Carlo, Rahela, Atko, Drago, Antoine, Sanja, Dejan, Halid, and others, created an unforgettable harmony. These are only some of the people working for a better future. It is their dream that the unity demonstrated by their beautiful harmonies will become the start of harmony and unity for all of Bosnia and Herzegovina.

The old bridge of Mostar is gone, destroyed during the fighting in the 1990s. It is now being rebuilt. History cannot be changed, but the future can. The people of Bosnia and Herzegovina are once again on a korzo. The world is watching, with hopeful anticipation, for the results of this stroll into Bosnia and Herzegovina's future.

Fact at a Glance

Country Name	Bosnia and Herzegovina
Location	Southeast Europe, bordering the Adriatic Sea and Croatia
Capital	Sarajevo
Area	19,741 square miles (51,129 square kilometers)
Land Features	Mountains and valleys; lowest elevation—sea level; highest elevation—Mt. Maglic (7,831 feet, 2,387 meters)
Climate	Hot summers and cold winters; areas of high elevation have short, cool summers and long, severe winters; mild, rainy winters along coast
Major Water Features	Sava River; 13-mile (21-kilometer) Adriatic Coastline
Natural Hazards	Earthquakes, forest fires, floods
Land Use	Arable land: 10% Permanent crops: 3% Other: 87%
Environmental Issues	Air pollution; limited urban waste disposal sites; water shortages and destruction of infrastructure due to the 1992–1995 civil strife
Population	Approximately 4,000,000 (July 2003 estimate)
Population Growth Rate	0.2% per year
Total Fertility Rate	1.3 (average number of children born to each woman during childbearing years)
Life Expectancy at Birth	Total population: 72 years Female: 75 years Male: 69 years
Ethnic Groups	Serb: 37.1%; Bosniak: 48%; Croat: 14.3%; Other: 0.5% (2003 estimate)
Religions	Muslim: 40%; Orthodox: 31%; Roman Catholic: 15%; Protestant: 4%; Other: 10%
Languages	Croatian, Serbian, Bosnian
Type of Government	Emerging Federal Democratic Republic
Executive Branch	Three-member rotating presidency
Independence	March 1, 1992

Administrative Divisions	Two administrative divisions called entities (Federation of Bosnia and Herzegovina and the Republic of Srpska), and an internationally supervised district in Brcko.
Currency	Convertible mark
Industries	Steel, coal, iron ore, lead, zinc, manganese, bauxite, vehicle assembly, textiles, tobacco products, wooden furniture, tank and aircraft assembly, domestic appliances, oil refining
Unemployment Rate	40% (2001 estimate)
Primary Exports	Miscellaneous manufactured goods, crude materials
Export Partners	Croatia, Switzerland, Italy, Germany
Imports	Machinery and transport equipment, industrial products, foodstuffs
Import Partners	Croatia, Slovenia, Italy, Germany
Transportation	Railroad: 634 miles (1,021 kilometers); Highways (total): 13,574 miles (21,846 kilometers); Highways (paved): 8,711 miles (14,020 kilometers); Airports: 27
Ports and Harbors	Bosanska Gradiska, Bosanski Brod, Bosanski Samac, and Brcko (all inland waterway ports on the Sava River), Orasje

History at a Glance

200,000 B.C.	Evidence of first human presence
4000 B.C.	Agriculture spreads across the Balkans
Third century B.C.	Illyricum Kingdom established
Second-first centuries B.C.	Romans conquer Illyricum
6 A.D.	Illyrians failed attempt to revolt against Romans
395	Roman Empire divided into Eastern and Western Empires
1463	Ottoman Turks conquer Bosnia
1875–1876	Balkan revolts against Turkish rule
1878	Council of Berlin ends Russo-Turkish War, Austria-Hungary occupies and governs Bosnia and Herzegovina
1882	Triple Alliance formed by Germany, Austria-Hungary, and Italy
1904	United Kingdom joins France and Russia to form the Triple Entente
1908	Bosnia and Herzegovina annexed into the Austro-Hungarian Empire
1914	Archduke Franz Ferdinand assassinated in Sarajevo; World War I begins
1918	World War I ends; Kingdom of Serbs, Croats, and Slovenes established
1929	Country's name changed to Yugoslavia
1941	Bosnia and Herzegovina invaded by Germany during World War II
1945	Bosnia and Herzegovina becomes one of six republics in Yugoslavia after the war; Tito gains leadership in Yugoslavia
1961	Non-Aligned Movement started by Marshall Tito, Gamal Abdel Nasser, and Jawaharlal Nehru
1984	Sarajevo hosts the fourteenth Winter Olympics
1989	Slobodan Milosevic elected President of Serbia; campaign of Serb nationalism begins
1990	Yugoslavia holds free elections; Serb and Croat nationalist parties win major victories over the Communist Party

1990	Serbia invades communities in Croatia; Bosnia and Herzegovina begins independence process
1991	Croatia retaliates against Serbs; ethnic fighting broadens to all groups; Bosnia and Herzegovina formally declares independence; Alija Izetbegovic elected President; siege of Sarajevo begins
1995	Siege of Sarajevo and fighting end; NATO troops arrive; Dayton Accord established
1996	Izetbegovic reelected President
2003	Sarajevo fails in its bid to host the 2010 Winter Olympics

Further Reading

Ali, Rabia, and Lawrence Lifschultz (eds). *Why Bosnia? Writings on the Balkan War*. Stony Creek, Conn.: The Pamphleteer's Press, 1993.

Gerolymatos, Andre. *The Balkan Wars: Conquest, Revolution, and Retribution from the Ottoman Era to the Twentieth Century and Beyond*. New York: Basic Books, 2002.

Gjelten, Tom. *Sarajevo Daily: A City and Its Newspaper Under Siege*. New York: HarperCollins Publishers, 1995.

Glenny, Misha. *The Balkans: Nationalism, War and the Great Powers, 1804-1999*. New York: Penguin Group, 2000.

Hukanovic, Rezak. *The Tenth Circle of Hell: A Memoir of Life in the Death Camps of Bosnia*. New York: Basic Books, 1996.

Judah, Tim. The Serbs: *History, Myth, and the Destruction of Yugoslavia*. New Haven, Conn.: Yale University Press, 1997.

Kaplan, Robert D. *Balkan Ghosts: A Journey Through History*. New York: Vintage Books, 1996.

Karahasan, Dzevad. *Sarajevo: Exodus of a City*. New York: Kodansha America, Inc., 1994.

Lovrenovic, Ivan. *Bosnia: A Cultural History*. London: The Bosnian Institute/Saqi Books, 2001.

Maass, Peter. *Love Thy Neighbor: A Story of War*. New York: Vintage Books,1996.

Magas, Branka and Ivo Zanic (eds). *The War in Croatia and Bosnia and Herzegovina 1991-1995*. London: The Bosnian Institute/Frank Cass, 2001.

Rieff, David. *Slaughterhouse: Bosnia and the Failure of the West*. New York: Touchstone, 1996.

Silber, Laura, and Allan Little. *Yugoslavia: Death of a Nation*. New York: Penguin Books, 1997.

Stoianovich, Traian. *Balkan Worlds: The First and Last Europe*. New York: M.E. Sharpe, Inc., 1994.

Tanter, Raymond, and John Psarouthakis. *Balancing in the Balkans*. New York: St. Martin's Press, 1999.

Thompson, Mark. *A Paper House: The Ending of Yugoslavia*. New York: Random House, Inc., 1992.

Winchester, Simon. *The Fracture Zone: A Return to the Balkans*. New York: HarperCollins Publishers, Inc., 1999.

Index

Adriatic Sea, 13, 16, 18, 23, 25, 30, 35, 36, 96
agriculture, 13, 16, 22, 24, 26, 36, 77, 89, 97
air pollution, 31
air transportation, 79, 82, 92
Albanians, 35
Allah, 57
alluvial basins, 21-22
Alsace and Lorraine region, 37
animal life, 16, 25, 27-28, 31
Annan, Kofi, 91
area, 12
Austro-Hungarian Empire, 90
    See also Habsburg Empire/Monarchy
Axis powers, 59

Balkan Mountains, 18
Balkan Peninsula, 17-18, 25, 32, 34, 35
Banja Luka, 13, 29, 82, 84-86, 93
birds, 28
Bjelasnica Range, 27
Black Hand, 38, 62
black market, 74
borders, 18, 20, 21, 29, 59
Bosanska Gradiska, 79
Bosanski Brod, 79
Bosanski Samac, 79
Bosna River, 29, 45, 46, 89
Bosnia, 35-36
    in Yugoslavia, 58
    See also Bosnia and Herzegovina
Bosnia and Herzegovina
    birth of, 9, 45, 48, 58
    and independence, 44-45, 58, 62
Bosniaks. See Muslims/*Bosniaks*
Bosnian language, 15, 54
Bosnian pizza, 76
Bosnian pot, 76
Bosnians, 55
Brcko, 47, 79, 84, 86, 87-89, 93
bridges, 8-9, 33
    Stari Most, 87, 101
"Brotherhood and Unity," 40

Broz, Josip, 40
    See also Tito, Marshall
Bulgaria, 1836
Byzantine Empire, 34, 61

Cabrinovic, Nedjelko, 38
cantons, 70
capital city. *See* Sarajevo
Castra, Fort, 86
Catholics, 9, 53, 60, 61
    See also Croats
cave bear, 27
caves. *See* karst features
cell phones, 80, 96-97
Center for Civic Education, 98
Children's Cemetery, 11
chopska salad, 76
Christmas, 60
cigarette smoking, 54
cities, 13, 18, 21, 29, 30, 31, 36, 47, 56, 79, 84-93
    See also Sarajevo
citizens
    informed, 96
    rights of, 70-72
civil war, 9
    and Banja Luka, 86, 93
    and Brcko, 87-89
    and Croats, 11-12, 14, 44, 45, 47, 51, 58, 60-61, 87, 89-90, 97
    and Dayton Accord, 15, 47-48, 64-65, 86, 87-88
    deaths in, 11-12, 14, 43, 45, 47, 53, 87, 94, 97
    and displaced people, 14, 47, 50, 71-72, 89, 90, 93, 100-101
    and economy, 74-76, 77-78, 79, 80-81
    and ethnic cleansing, 14, 46-47, 50, 89-90, 96, 97, 100
    and foreign aid, 74-76, 82, 92, 95
    and forests, 27
    and future, 95-100
    and hills, 9, 11, 45
    and Jews, 51

107

Index

kindness during, 50-51, 93
and land mines, 27, 31, 46, 79, 91
legacy of, 14-15, 51
and Milosevic, 44-47, 48, 51
and Mostar, 86-87, 101
and Muslims/*Bosniaks,* 11-12, 14, 44-45, 47, 50-51, 58, 60-61, 87, 89-90, 97
and nationalism, 43, 44-47, 52, 58, 97
and NATO, 47
and postwar building efforts, 49-51
and regionalism, 43, 44, 52
and religion, 43, 44, 52
and Sarajevo, 11-12, 32, 45-47, 80-81, 90, 91
and Sarajevo rose, 32, 51
and Serbs, 9-12, 14, 44-47, 50-51, 58, 60-61, 89-90, 97
and Travnik, 89-90, 93
Civitas Bosnia and Herzegovina, 98-99
Clinton, Bill, 47, 91
coastline, 13, 16, 18, 30, 96
coffee, 14
Cold War, 40-41
communications, 80-81, 82
Communist rule, 15
and forests, 27
and free elections (1990), 44
and Tito, 40-41, 42, 43, 44, 52, 57-58, 59, 100
and World War II, 40
computers, 80
Congress of Berlin, 36, 37, 62
constitution, 15, 62, 64-72, 73
Constitutional Court, 67, 69
convertible marka, 82
Council of Europe, 71
Council of Ministers, 69
Croat language, 54
Croatia, 12, 13, 14, 18, 20, 59
Bosnia and Herzegovina incorporated into, 40
and civil war, 60
and Croats, 59-60

history of, 35, 58-59, 61
and independence, 9, 59, 62
and Milosevic, 44
as part of Yugoslavia, 44, 59
and Tudjman, 58, 66
and World War II, 40, 59
See also Croats
Croatian Democratic Community of Bosnia and Herzegovina, 60
Croatian Kingdom, 58-59
Croatian language, 15
Croatian Peasants' Party of Bosnia and Herzegovina, 60
Croats, 52, 58-60, 63, 64
and Brcko, 88-89
as Catholics, 9, 53, 60, 61
and civil war, 11-12, 14, 44, 45, 47, 51, 58, 60-61, 87, 89-90, 97
in Croatia, 44
and Federation of Croats and Muslims, 47, 48
and government, 60, 62, 64-69
history of, 35, 37, 58-59, 61
and Kingdom of Serbs, Croats, and Slovenes, 40, 62
and language, 54
and Mostar, 86, 87
and nationalism, 58, 62, 66-67
population of, 14, 53, 58
Serbs and Muslims/*Bosniaks versus,* 57, 58, 59. *See also* and civil war, *above*
and Tito, 43
and union of Slovenes and Serbs, 40
and World War II, 59
See also Croatia
culture, 8, 34, 35, 84
See also ethnic groups
currency, 66, 74, 82
Cyrillic alphabet, 54

Dalmatia, 21, 35
Dayton Accord, 15, 47-48, 64-65, 86, 87-88
democracy, 51, 98-99

Index

Diana, Princess, 91
Dinaric Alps, 18, 20, 21-22, 23, 25
displaced people, 14, 47, 50, 71-72, 89, 90, 93, 100-101
diversity. *See* ethnic groups
Drina River, 18, 28, 29, 30
Duklja, 35

earthquakes, 20
Easter, 60
Eastern Roman Empire, 35
economy, 14, 15, 28, 51, 74-83, 86, 92, 96-97
education, 14, 86, 97, 98-99
eel fisheries, 28
Eid al-Adha, 57
Eid al-Fitr, 57
elevation, 22, 23, 25
  See also mountains
employment, 14
energy resources, 30, 78, 86
Entente, 37
entities, 13, 47, 48, 50-51, 67, 69-70
  See also Federation of Bosnia and Herzegovina; Republic of Srpska
environmental hazards, 20, 30
environmental issues, 27, 28, 30-31
ethnic cleansing, 14, 46-47, 50, 89-90, 96, 97, 100
ethnic groups, 9, 13, 18, 52
  and Banja Luka, 86
  and Brcko, 87-89
  and Dayton Accord, 48
  and demise of Yugoslavia, 9
  and economy, 82
  and foreign investment, 97
  and future, 94-95, 97-101
  and government, 15, 58, 60, 62, 64-69, 69-72, 72
  and Habsburgs, 37
  and intermarriage, 14
  and languages, 14-15, 35, 54
  and Mostar, 87
  and Sarajevo, 14, 45, 92-93, 100
  and tolerance, 63, 96, 100
  and Travnik, 89-90
  See also civil war; Croats; Muslims/*Bosniaks*; Serbs
euro, 82
European Court of Justice, 69
European Union, 44, 72
exports, 81

falconry, 28
Federation of Bosnia and Herzegovina, 13, 56, 60, 61, 67, 68, 69, 70, 76, 87-88
Federation of Croats and Muslims, 47, 48
Ferdinand, Archduke Franz, 9, 38, 62, 90
Ferhadija Mosque, 86
fish/fishing, 28, 31
flag, 12
foods, 14, 28, 76, 78, 81, 89
foreign aid, 72, 74-76, 82, 92, 95, 96
foreign investment, 77, 78, 81-83, 97
foreign policy, 12-13, 68, 72, 74-76, 91-92
forests, 13, 21-22, 26-27, 31, 78
fossil fuels, 78
France, and World War I, 37, 38
future, 94-101

geology, 20
Germanic tribes, 35
Germany
  and World War I, 37-38
  and World War II, 15, 34, 40, 59
glaciers, 28
Glamoc, 21
Goths, 35
government, 51, 64-73
  and constitution, 15, 62, 64-72, 73
  and entities, 13, 47, 48, 50-51, 67, 69-70. *See also* Federation of Bosnia and Herzegovina; Republic of Srpska
  and ethnic groups, 15, 58, 60, 62, 64-69
Gypsies, 52, 68

109

Index

Habsburg Empire/Monarchy, 18, 34, 36, 37, 62, 90
Hajj, 57
Herzegovina, 13, 35
 See also Bosnia and Herzegovina
Highland conditions, 23, 25
hills, 8, 9, 11, 13, 20, 21-22, 45
history, 15, 18, 32-41, 86
Hodidjed, 90
hospitality, 14
House of Peoples, 67-68
House of Representatives, 67, 68
Hrobatos, Chief, 58
Humid Subtropical climate, 23
Hungarians, 15, 35, 59, 61
hunting, 28
hydroelectric power, 30, 78, 86

Ice Age, 28
Illyrians, 34-35, 58, 86
Implementation Forces (IFOR), 47, 74-75, 95
imports, 81
income levels, 76
independence
 of Bosnia and Herzegovina, 44-45, 58, 62
 of Croatia, 9, 59, 62
 of Macedonia, 62
 of Slovenia, 62
inflation, 82
international agreements, 72
International Monetary Fund, 72
Internet, 80, 97
Interpol, 72
Islam. *See* Muslims/*Bosniaks*
Islamic calendar, 57
Italy, and Triple Alliance, 37
Izetbegovic, Alija, 45, 48, 58

Jews, 51, 52, 54, 63, 67-68
Julian calendar, 60

Kaaba, 57
Kallay, Benjamin, 37

Karadzic, Radovan, 48
karst features, 20, 22, 26, 28-29, 30, 31
Kastel, 86
Kingdom of Serbs, Croats, and Slovenes, 40, 62
Koran (Qur'an), 56-57
korzo (stroll), 8, 101
Kosovo, 62

lakes, 28
land features, 16, 18, 20-31
land mines, 27, 31, 46, 79, 91
languages, 14-15, 35, 54
latitude, 17, 22
leech gathering, 28
Liberal Bosniak Party, 58
lichens, 26
life expectancy, 54
livestock, 22
Livno, 21
location, 15, 16, 17-18, 20, 22-23, 32-34
lowlands, 22

Macedonia, 9, 44, 62
Maglic, Mount, 21
manufacturing, 14, 77-78, 81, 97
markets, 76, 92
Mecca, 57
Mediterranean climate, 23-24
metallurgical plants, 31
Middle Ages, 35-36
Miljacka River Valley, 45, 46, 90
Milosevic, Slobodan, 44-47, 48, 51, 58, 62, 66
minerals, 20, 30, 78
Mohammed, 57
Montenegro, 36, 40, 44
moor, 21-22
mosses, 26
Mostar, 13, 21, 36, 47, 56, 84, 86-87, 93
mountains, 13, 18, 20, 21-22, 23, 25, 26, 27, 28, 45
Muslim-Bosniak Party, 58

110

Index

Muslims/*Bosniaks,* 9, 52, 55-58, 63, 64-69
   beliefs of, 56-57
   and Bosnia and Herzegovina, 44
   and Brcko, 88-89
   and civil war, 11-12, 14, 44-45, 47, 50-51, 58, 60-61, 87, 89-90, 97
   and ethnic minority status, 55
   and Federation of Croats and Muslims, 47, 48
   and Izetbegovic as president, 45, 48, 58
   and language, 54
   and Mostar, 86, 87
   and Ottoman Turks, 28, 36-37, 55, 61
   population of, 14, 53, 55
   and religion, 56-57
   Serbs and Croats *versus,* 57, 58, 86.
      See also and civil war, *above*
   and Tito, 43

Nasser, Gamal Abdel, 40
national government, 66-69
nationalism, 9
   and civil war, 43, 44-47, 52, 58, 97
   and Croats, 58, 62, 66-67
   and demise of Yugoslavia, 9
   and Habsburgs, 37
   and Kingdom of Serbs, Croats, and Slovenes, 40
   and Ottoman Turks, 36
   and Serbs, 58, 61-62, 66, 67, 86
   and Tito, 41, 43
   See also civil war
NATO (North Atlantic Treaty Organization), 47, 71
natural resources, 13, 20, 21-22, 26-27, 30, 31, 78
Nazis, 15, 40
   See also Germany
Nehru, Jawaharlal, 40
Neretva River, 22, 28, 30
Neretva River Valley, 24, 86

Neum, 18, 30
New Croatian Initiative, 60
New Year's Day, 60
Non-Aligned Movement, 40-41, 42, 57-58
non-governmental organizations (NGOs), 75-76, 92, 98-99

Office of the High Representative (OHR), 86
Olympics (Winter)
   1984, 11, 16, 42, 43, 90-91, 92, 96
   2010, 91, 101
Orasje, 79
Orthodox Christians, 9, 53, 60, 61
   See also Serbs
Oslobodjenje, 80-81
Ottoman Turks, 15, 18, 28, 34, 36-37, 55, 59, 61-62, 86-87, 89

Paleolithic Age, 34
Parliamentary Assembly, 67, 69
people, 13-14
   See also ethnic groups
physical geography, 8, 9, 11, 12, 16-18, 20-31, 80
physical landscape, 13
plains, 16, 21, 22, 23
plant life, 16, 25, 26-27
plateaus, 20, 21-22
political parties, 58, 60, 62, 72-73
population, 12, 14, 22, 53-54, 55
ports, 35, 79, 89, 96
postal system, 66, 81, 97
Powell, Colin, 92
precipitation, 13, 23, 24, 25
presidency, 67, 68-69
Princip, Gavrilo, 38, 62
Project Citizen, 98-99
Protestants, 54, 63
Pula, 33

radio, 80
railroads, 79, 80, 86
Ramadan, 57

Index

regionalism
   and civil war, 43, 44, 52
   and Tito, 43
religion, 9, 51, 52, 53-54
   and civil war, 43, 44, 52
   and demise of Yugoslavia, 9
   and Habsburgs, 37
   and intermarriage, 14
   and Kingdom of Serbs, Croats, and Slovenes, 40
   and Sarajevo, 92-93
   and Tito, 41, 43, 57-58
   See also Catholics; Jews; Muslims/*Bosniaks*; Orthodox Christians
reptiles, 28
Republic of Srpska, 13, 47, 48, 50-51, 59, 60, 65, 67, 68, 69, 70, 76, 81, 84, 87-88
restaurants, 76
river valleys, 16, 24, 45, 46, 85-86, 90
rivers, 16, 18, 22, 28, 29-30, 31, 45, 46, 79, 86, 87, 89
roads, 33, 80, 86
Roman Empire, 15, 30, 33, 34-35, 86
rugged land, 20-22
Russia
   and Russo-Turkish War, 36, 62
   and Slavs, 37-38
   and World War I, 37-38
Russo-Turkish War, 36, 62

Sana River, 29
Sanja, 50-51
Sarajevo, 21, 45, 90-93
   and assassination of archduke, 9, 38, 39, 90
   as capital, 13, 67, 90, 92
   and civil war, 11-12, 32, 45-47, 80-81, 90, 91
   and Dayton Accord, 86
   deaths in, 11-12
   and economy, 79, 82
   and ethnic groups, 14, 45, 92-93, 100
   and foreign relations, 91-92
   history of, 90
   and *korzo* (stroll), 8, 101
   Muslims/*Bosniaks* in, 56
   and newspapers, 80-81
   population of, 13, 84, 90
   and rose, 32, 51
   springs near, 30
   and Stari Grad, 78, 92
   and transportation, 92
   and Winter Olympics (1984), 11, 16, 42, 43, 90-91, 92, 96
   and Winter Olympics (2010), 91, 101
   and World War I, 8-9, 90
Sarajevo Airport, 79, 82
satellite technology, 80
Sava River, 18, 22, 28, 29-30, 79, 87, 89
sea travel, 13
Serb Civic Council, 62
Serb Democratic Party, 62
Serb language, 54
Serb Radical Party, 62
Serbia, 35, 36, 37
   and Kingdom of Serbs, Croats, and Slovenes, 40
   and Milosevic, 44-47, 48, 51, 58, 62, 66
   and Yugoslavia, 44
   See also Serbs
Serbia and Montenegro, 9, 12, 13, 14, 18, 21, 29, 61, 62
Serbian language, 15
Serbo-Croatian language, 14-15, 35, 54
Serbs, 52, 60-63, 64
   and assassination of archduke, 9, 38, 62, 90
   and Brcko, 88-89
   and civil war, 9-12, 14, 44-47, 50-51, 58, 60-61, 89-90, 97
   and Cyrillic alphabet, 54
   and government, 62, 64-69
   and Habsburgs, 37

112

Index

history of, 35, 61-62
and Kingdom of Serbs, Croats, and Slovenes, 40, 62
and language, 54
Muslims/*Bosniaks versus,* 57, 58, 86. *See also* civil war, *above*
and nationalism, 58, 61-62, 66, 67, 86
as Orthodox Christians, 9, 53, 60, 61
and Ottoman Turks, 61-62
population of, 14, 53, 60
in Serbia, 44
and Tito, 43
and Union of Slovenes and Croats, 40
and World War I, 9, 38, 62, 90
Yugoslavia led by, 44, 45
See also Republic of Srpska; Serbia
shape, 12
skiing, 24
Slavs, 35, 37-38, 86
See also Croats; Serbs; Slovenes
Slovenes
and government, 62
history of, 35
and Kingdom of Serbs, Croats, and Slovenes, 40, 62
and union of Croats and Serbs, 40
Slovenia, 9, 44, 62
small businesses, 97
Sniper's Alley, 11
snow, 13, 24, 25
soils, 22
Sophie, the Duchess of Hohenberg, 9, 38, 90
Southeast Europeans, 18
Southern Alps, 20
Soviet Union, 34, 40-41
sports, 24, 28
springs, 30
Stabilization Force (SFOR), 47, 75, 92, 95
Stari Grad, 78, 92
Stari Most bridge, 87, 101

Starigrad, 32
steel, 78
streams, 28-29, 30

tarns, 28
television, 80
timber industry, 13, 21-22, 26-27, 31, 78
Tito, Marshall, 40-41, 42, 43, 44, 52, 57-58, 59, 100
tolerance, and ethnic groups, 63, 96, 100
Tomislav, King, 58
tourism, 14, 30, 80
trade, 13, 15, 35, 79-80, 81-82, 87, 89, 96
transportation, 13, 20-21, 33, 79-80, 81, 82, 86, 92
Travnik, 21, 36, 84, 89-90, 93
Triple Alliance, 37-38, 39
Triple Entente, 38, 39
Tudjman, Franjo, 58, 66
Turks. *See* Ottoman Turks
Tuzla, 30, 86
Tvrtko, Ban Stefan, 35-36

Una River, 18, 29
unemployment, 76, 77, 78, 97
UNESCO (United Nations Educational, Scientific, and Cultural Organization), 72
union of Slovenes, Croats, and Serbs, 40
United Kingdom, and Triple Entente, 38
United Nations, 45-46, 72, 88
University of Banja Luka, 86
Ustachi, 40

valleys, 16, 21, 24, 45, 46, 85-86, 90
Visigoths, 35
Vrbanja River, 86
Vrbas River, 29, 30, 86
Vrbas River Valley, 85-86
Vucko, 90

Index

waste disposal, 31
water features, 13, 16, 22, 23, 25, 28-30, 31, 45, 46, 79, 86, 87, 89
water supply, 31
water transportation, 79
weather/climate, 13, 16, 22-25
winds, 23, 25
Winter Olympics. *See* Olympics
woodland, 27
World Bank, 71
World Health Organization, 72
World War I, 8-9, 37-39, 62, 90, 96
World War II, 40, 57, 59, 92-93
written language, 54

Yugoslavia, 18, 33
   disintegration of, 9, 42, 58, 62. *See also* civil war
   and free elections (1990), 44
   Kingdom of Serbs, Croats, and Slovenes as, 40, 62
   six republics in, 40, 44, 58, 59
   and Tito, 40-41, 42, 43, 44, 52, 57-58, 59, 100

Picture Credits

page:

- 10: © Lucidity Information Design, LLC
- 11: Courtesy of Douglas Phillips
- 12: Courtesy of Douglas Phillips
- 17: © Michael S. Yamashita/CORBIS
- 19: © Lucidity Information Design, LLC
- 21: © Alain Le Garsmeur/CORBIS
- 26: © Otto Lang/CORBIS
- 29: © Otto Lang/CORBIS
- 33: Courtesy of Douglas Phillips
- 39: Associated Press, AP
- 43: Associated Press, AP
- 46: Courtesy of Douglas Phillips
- 48: © Peter Turnley/CORBIS
- 49: Courtesy of the Library of Congress Geography and Map Division
- 50: Associated Press, AP
- 53: Associated Press, AP
- 56: Associated Press, AP
- 61: Associated Press, AP
- 65: © Reuters NewMedia Inc./CORBIS
- 66: Courtesy of Douglas Phillips
- 75: © Jon Hicks/CORBIS
- 85: Associated Press, AP
- 88: © Otto Lang/CORBIS
- 91: © Dean Conger/CORBIS
- 95: Associated Press, AP
- 99: Courtesy of Douglas Phillips

Cover: © Roman Soumar/CORBIS

About the Author

DOUGLAS A. PHILLIPS is a lifetime educator and writer who has worked and traveled around the world to over 75 nations. He has traveled to Bosnia-Herzegovina over a dozen times in his work to help develop civic education in that country. He loves the country and the people and has had the good fortune of working with hundreds of wonderful Croats, Serbs, Bosniacs, Jews, and others in this war-torn nation. During his career he has worked as a middle school teacher, as a curriculum developer, writer, and as a trainer of educators in various locations around the world. He has served as the President of the National Council for Geographic Education and has received the Outstanding Service Award from the National Council for the Social Studies along with numerous other awards. He, his wife Marlene, and their three children, Chris, Angela, and Daniel have lived in South Dakota and Alaska, but he and his family now reside in Arizona where he writes and serves as an educational consultant.

CHARLES F. ("FRITZ") GRITZNER is Distinguished Professor of Geography at South Dakota University in Brookings. He is now in his fifth decade of college teaching and research. During his career, he has taught more than 60 different courses, spanning the fields of physical, cultural, and regional geography. In addition to his teaching, he enjoys writing, working with teachers, and sharing his love for geography with students. As consulting editor for the MODERN WORLD NATIONS series, he has a wonderful opportunity to combine each of these "hobbies." Fritz has served as both President and Executive Director of the National Council for Geographic Education and has received the Council's highest honor, the George J. Miller Award for Distinguished Service.